Innovation and Adaptation in War

The Belfer Center Studies in International Security book series is edited at the Belfer Center for Science and International Affairs at Harvard Kennedy School and is published by the MIT Press. The series publishes books on contemporary issues in international security policy, as well as their conceptual and historical foundations. Topics of particular interest to the series include the spread of weapons of mass destruction, internal conflict, the international effects of democracy and democratization, and U.S. defense policy.

A complete list of Belfer Center Studies in International Security books appears at the back of this volume.

Innovation and Adaptation in War

Matthew A. Tattar

Belfer Center Studies in International Security
The MIT Press
Cambridge, Massachusetts
London, England

This book was set in Minion Pro by Jen Jackowitz.
Printed and bound in the United States of America.

Library of Congress Cataloging-in-Publication Data
Names: Tattar, Matthew A., author.
Title: Innovation and adaptation in war / Matthew A. Tattar.
Description: Cambridge, Massachusetts : The MIT Press, 2024. | Series: Belfer Center studies in international security | Includes bibliographical references and index.
Identifiers: LCCN 2024023047 (print) | LCCN 2024023048 (ebook) | ISBN 9780262544856 (paperback) | ISBN 9780262372718 (pdf) | ISBN 9780262372725 (epub)
Subjects: LCSH: Military doctrine—Case studies. | Armed forces—Organization—Case studies. | Naval art and science—Case studies. | Military history, Modern—20th century—Case studies.
Classification: LCC U21.2 .T364 2024 (print) | LCC U21.2 (ebook) | DDC 355.4—dc23/eng/20240828
LC record available at https://lccn.loc.gov/2024023047
LC ebook record available at https://lccn.loc.gov/2024023048

10 9 8 7 6 5 4 3 2

Contents

List of Abbreviations

AAA	Anti-Aircraft Artillery
ACM	Air Combat Maneuvering
AEW	Airborne Early Warning
AFSCC	Air Force Special Communications Center
ARM	Anti-Radiation Missile
ASCM	Anti-Ship Cruise Missile
ASD	Anti-Submarine Division
ASW	Anti-Submarine Warfare
BRN	British Royal Navy
BuAer	Bureau of Aeronautics
BuPers	Bureau of Personnel
BuShips	Bureau of Ships
CAP	Combat Air Patrol
CAS	Close Air Support
CIWS	Close-In Weapon System
CNO	Chief of Naval Operations
COMINCH	Commander in Chief, U.S. Fleet
CRT	Cathode Ray Tube
CTF 69	Composite Task Force 69
DAMS	Defensive Armament of Merchant Ships
DMZ	Demilitarized Zone
DOS	Dedicated Organizational Subunit
ECM	Electronic Countermeasures
EWO	Electronic Warfare Officer
FBM	Fleet Ballistic Missile
GCI	Ground-Controlled Intercept
IAD	Integrated Air Defense
IFF	Identification Friend or Foe
IJA	Imperial Japanese Army
IJN	Imperial Japanese Navy
INS	Internal Navigation System
IT-RMA	Information Technology Revolution in Military Affairs
JCS	Joint Chiefs of Staff
LGB	Laser-Guided Bombs

MAD	Magnetic Anomaly Detector
MOE	Measures of Effectiveness
MTI	Moving Target Indicator
NATO	North Atlantic Treaty Organization
NVA	North Vietnamese Army
NVN	North Vietnamese Navy
QRC	Quick Reaction Capability
RAF	Royal Air Force
RAS	Replenishment at Sea
RFA	Royal Fleet Auxiliary
RHAWS	Radar Homing and Warning System
RMA	Revolution in Military Affairs
ROE	Rules of Engagement
RTAFB	Royal Thai Air Force Base
SAC	Strategic Air Command
SAM	Surface-to-Air Missile
SAS	Special Air Service
SEAD	Suppression of Enemy Air Defenses
SHAR	Sea Harrier
SIOP	Single Integrated Operational Plan
SLBM	Submarine-Launched Ballistic Missile
SLOC	Sea-Lanes of Communication
SSN	Nuclear Attack Submarine
STARM	Standard Anti-Radiation Missile
STUFT	Ships Taken Up from Trade
TAC	Tactical Air Command
TLRP	Top-Level Review Panel
USAF	U.S. Air Force
USN	U.S. Navy

List of Illustrations

Chapter 1

Chapter 2

Chapter 3

Chapter 4

Chapter 5

Chapter 6

Chapter 7

Acknowledgments

THIS BOOK HAS been a long time coming. It has been an exercise in persistence, but thankfully not a solitary one. I owe a huge debt of gratitude to my advisor at Brandeis University, Robert Art. He set the standard for how to be a thoughtful and compassionate mentor. My argument would also have been woefully incomplete without incorporating Harvey Sapolsky's early comments.

I must also thank those that have helped me nurture and sustain this book project. My time at the Belfer Center kick-started it, and Steven Miller graciously extended my fellowship to account for an unexpected stint of military service in Afghanistan. Mentors and colleagues at the center provided time and opportunity in an encouraging environment to develop a proposal and work on the manuscript. Previous editors of this series, Sean Lynn-Jones and Morgan Kaplan, were especially helpful in improving the work through the review process.

The world of "applied" research can be very rewarding, except in free intellectual time. I owe thanks to the Naval War College's Richard "Twig" LaBranche and Michael O'Hara for encouraging me to finish this project and providing the time to do so. Alexis Montouris Ciambotti provided invaluable research assistance while I transformed this manuscript into book shape.

Thank you also to the team at the MIT Press, and especially to Karen Motley at the Belfer Center.

I will always be grateful for the friendship and encouragement of "Productivity Corner" in the War Gaming Department: Heath "Hank" Brightman, Jon Scott Logel, Margaret Polski, and Joseph Mroszczyk.

Thank you to my parents, Donna and Terry, who patiently supported and encouraged me through the years. I am extremely lucky to have a mentor and accomplished academic in my life who also answers to "Dad."

Most important, I could not have gotten this book across the finish line without the love and support of my wife, Abhivyakti. From the bottom of my heart, thank you, Gauri.

Introduction

PEOPLE AND ORGANIZATIONS in many fields of human endeavor have long found that formulating new and more efficient ways of doing things—in other words, innovating—is a good thing. This is particularly true for competitive endeavors. Find a new way to produce goods, distribute information, or increase efficiency, especially in a way that is surprising or unexpected to one's competitor, and one can expect to have a "leg up" on the competition. This conventional logic holds true in areas such as athletics, business, and especially warfare.

But is this conventional logic actually correct in the realm of military innovation? Is it always a good thing to come up with a new mode of operation first? It is hard to argue that finding better ways to do things than one's opponent is a bad thing, but then again, the question is rarely as simple as choosing between innovating or not innovating. Military organizations must make tough choices. Where should the organization commit its finite, often scarce resources? For example, are force readiness or new weapon systems more important? Furthermore, why do some of the most innovative military organizations in history fail to capitalize on their advantages and eventually lose the wars they fight? Some military organizations are very good at doing some things and very incompetent at others. The Japanese and German militaries of the first half of the twentieth century come to mind as falling into this category. Does focusing "too much" on innovation impose enough of a cost to threaten, even slightly, the ability to achieve victory? Is innovation worth it, and if so, what kind? If we, as analysts of human conflict using the imperfect laboratory of human history, can hold other factors affecting victory or defeat in war relatively "constant," can we establish whether an innovation can provide that extra advantage that will lead to victory?

The answer to this puzzle lies in the interactive nature of warfare. Competition—with warfare being viewed as the most extreme competition of all—is fundamentally interactive. In general terms, when one side does something, the other side reacts in some way. Through its own action or inaction, the adversary who is the target of an innovation almost always has a say in how effective that innovation is. To examine wartime innovation strictly from the point of view of the innovator is to leave out half of the story. When an innovation makes an appearance on the battlefield, it is only decisive as long as the opponent fails to develop an effective response to it. Therefore, the analysis of an innovation's effects naturally shifts to how the opponent's adaptation to it progresses.

The crux of the argument presented in this book is that the decisive factor in successful adaptation, more often than not, is a well-positioned or flexible organization. In other words, how many component parts of an organization are affected by an innovation? If there is just one, it can take charge of the adaptation effort, acting as a clearinghouse for new ideas and directing the overall effort. If it is more than one, subunits within the adapting military might compete for the privilege of assuming responsibility for adaptation or, conversely, might shun it. It follows from this that if the adapting military has only one subunit responsible for an innovation, it avoids the whole question of internal competition over, or neglect of, the adaptation problem. The organizational literature has proven that competition within an organization is not generally helpful, especially when the competing subunits have different goals or conceptions of the organization's purpose.[1]

Resources are also important but less decisive for three main reasons. First, organizations that have plenty of resources are not always good adapters. Second, while military leadership has lamented resource shortages innumerable times in the past, organizational priorities and decisions have a decisive effect on the distribution of resources within a military service. Where the resources go can be just as important as whether or not they exist. An innovation may seem to pose a tough resource-shortage problem for an adapter because there might be very little to spare for the task of adaptation. A deeper examination, however, often reveals that this problem is actually due to reluctance on the part of the adapter to divert resources from a favored combat arm or mode of operation. Thus, a resource problem often springs from organizational causes. Third, simply committing more resources to blunt an innovation can be incredibly wasteful and costly in terms of lives lost in battle.

With this relationship between organization and resources in mind, the independent variable of this study is the ability to organize to meet the new threat. The innovation imposes the need to develop a policy response on the adapting organization. The nature and target of the innovation determine what part or parts—which I call "subunits"—of the adapting organization must respond in some way. Adapting militaries have four organizational responses available to them in terms of subunits. First, they can have multiple (two or more) subunits that have responsibility for adaptation. This in turn leads one of two possible outcomes. The subunits may compete with one another to take the lead in adaptation, or both may shun responsibility because they assess it as outside their primary role or purpose. Either of these results is inefficient because the subunits will waste effort by competing with one another, or make no effort to solve the problem. Second, the organization may not designate or create any subunit dedicated to addressing the problem posed by the innovation. With no units specifically dedicated to solving the problem, the organization as a whole will find it very difficult to find an adequate solution. Third, the organization can form a review panel or special commission to study the problem and evaluate various responses, and then parcel out the necessary changes to relevant parts of the organization. This *ad hoc* or temporary arrangement enables successful adaptation, but is vulnerable to inefficiencies in development and implementation of the new methods of warfare. Fourth, the organization can form one subunit with the mandate and expertise to formulate a response to the innovation. This response is most likely to generate efficient adaptation because of the undivided attention that the subunit is able to commit to the problem. It does not have to compete with other subunits, and it will not ignore the problem because the problem is the focus of its existence.

The case studies in this volume demonstrate that the ability to form a dedicated subunit or, to a lesser extent, a top-level panel or conference in order to study the problem and formulate a solution, is predictive of adapting capability. Available resources are necessary for adaptation, but not sufficient. In turn, this says important things about how decisive innovation can be. If an innovation is launched against an organizationally flexible opponent, its decisive effect is likely to be limited in scope and duration. Conversely, if the adapting organization is poorly placed and inflexible, the innovation is likely to be more decisive.

Case studies with a qualitative-method focus are the most appropriate for this analysis, because this subject defies quantification. Also, it is necessary to open up the black box of organizational behavior during the rare event of major war involving Great Powers. This is especially significant because the researcher cannot rely upon large statistical samples to establish causality or correlation between variables. One of the frequent shortcomings of qualitative research in comparison to more formal methods is that many of the processes under examination are underspecified. What exactly is the process, and how does examination of it reveal causation, as opposed to just correlation? Historical details and narrative must fit within a structure. Inclusion of historical detail for its own sake is rarely informative. By identifying the specific internal organizational processes that will be under examination, then finding them at work in the adapting organization and examining the nuances and interplay in the innovator-adapter relationship in the various cases, this book addresses and corrects these potential shortcomings.[2]

The level of analysis is also important. Tactics of small units in the modern era can change with a frequency that can confound meaningful analysis. Strategic-level decisions depend largely on conditions particular to the country making them. The operational level of analysis, or the behavior of combat arms of a particular service and their interaction with one another, is the most illuminating level at which to assess the effects of innovation. Where different navies fight (i.e., strategy) depends on which country they serve. Individual units can change how they operate (i.e., tactics) relatively frequently. The different types of vessels in all navies, however, need to work together at the operational level in some way.[3] Thus, the central puzzle I seek to clarify with this volume is whether or not innovation is operationally decisive in war.

First, this book will provide the theoretical underpinnings of the argument. Historical lessons and organizational theory are important for this task. Modern military organizations continually face hard choices between maintaining force readiness and developing new ways to accomplish their missions. This project aims to better inform those decisions by evaluating whether or not innovation has proven to be decisive. Incidentally to this process, it demonstrates that organizations that pursue avenues of innovation too far before a conflict often limit their ability to adapt once a conflict begins. The innovators in these cases have often locked themselves into a certain policy path that makes it difficult to react to changing situations imposed by the adapters.

The first case under examination is that of the British Royal Navy (BRN) in World War I. Prepared to fight a decisive fleet engagement with the Imperial German Navy, it instead faced an unexpectedly severe threat to its commerce by German U-boats. This novel way of attacking commerce caught the BRN organizationally flat-footed. Too many of its subunits had responsibility for commerce protection. Furthermore, besides being divided and uncoordinated, these subunits also did not have the assets with which to deal with the problem. Those assets, namely the bulk of the BRN's destroyers, were allocated to the battle fleet. The battle fleet was not inclined to strip its battleships of anti-submarine protection for a mission that was perceived as too "defensive" in an otherwise offensive-minded navy. Thus the most obvious solution to the U-boat problem, a comprehensive convoy system for all vulnerable shipping, failed to materialize.

Once the BRN created a subunit whose sole responsibilities were commerce protection and anti-submarine warfare, however, the idea of the convoy gained a dedicated advocate within the service. Once the BRN adopted the convoy system, the effectiveness of the U-boat innovation decreased significantly. The BRN's leadership consistently lamented its lack of anti-submarine warfare resources, but poor organization was more responsible for its lack of an adequate response to the U-boat.

World War II provides a host of examples of wartime innovation. As this work focuses primarily on navies, the Pacific campaigns are central to my analysis. The U.S. Navy (USN), specifically its large force of battleships, was the primary target of the Imperial Japanese Navy (IJN). The IJN was the first to develop a naval air-strike arm capable of sinking battleships. On the opening day of the Pacific War, Japan surprised the United States with its innovative use of carrier air power at Pearl Harbor. It attacked the USN at a point when battleships were still dominant in the Navy, but after carrier aviation had secured a permanent (if supporting) role within the fleet. The result of the Pearl Harbor attack was that rather than charge across the Pacific with its battle line as called for in War Plan Orange, the USN shifted its focus to carrier war.[4] The tragedy at Pearl Harbor created a streamlined organization for the USN, making its adaptation easier and its counterattack more effective. The U.S. adaptation was both more effective and more robust than the initial Japanese innovation.

The second portion of the innovation-adaptation relationship between the Japanese and U.S. battle fleets was the advent of the kamikaze. Japanese naval aviation's effectiveness had faced a precipitous decline, primarily due to the effectiveness of the U.S. adaptation and the length of the war. Suicide tactics—using manned aircraft as cruise missiles—immediately made Japanese aircraft much more deadly; however, the same adaptations that the USN undertook to make its carrier fleet more potent offensively and better protected defensively made it well-placed to face the new threat. Superior tactics, new technology, and huge numbers ultimately blunted kamikaze tactics.

The third case study examines the U.S. Air Force (USAF) and its adaptation to North Vietnam's integrated air defenses. In its prosecution of the Rolling Thunder campaign, the USAF was unprepared for North Vietnam's "people's war in the air," which made use of surface-to-air missiles and anti-aircraft artillery in ways that challenged the USAF's ability to complete its mission. Its organizational focus was nuclear war with the Soviet Union using strategic bombers. North Vietnam used Soviet technology, but employed it in unexpected ways. The split between Strategic Air Command (SAC) and Tactical Air Command (TAC), coupled with SAC's dominance within the USAF, meant that the mission of suppressing enemy air defense was neglected. Once losses began to mount, the USAF convened the Dempster Commission to examine the problem, and out of this *ad hoc* subunit came new electronics, weapons, and aircraft to combat the problem. This case demonstrates that a top-level review panel can serve many of the same adaptation functions as a dedicated subunit, as these adaptations steadily degraded the surface-to-air missiles' effectiveness for the remainder of Rolling Thunder and also during the Linebacker operations of 1972.[5]

The fourth case study returns to adaptation by navies by again examining the BRN, this time during the Falklands War of 1982. This unexpected conflict saw the BRN well on its way in transitioning from its "East of Suez," blue-water focus, to that of anti-submarine warfare against the Soviet Union. The fleet that sailed to reverse the Argentinian occupation of the Falklands was thus dependent upon the British Royal Air Force (RAF) for air cover, yet it was thousands of miles from the nearest RAF base. Also, the vast majority of its anti-aircraft assets could not engage modern low-level targets. Thus, the Argentinian innovation, a combination of Exocet sea-skimming anti-ship missiles and extremely low-level air attacks, presented the BRN with a threat that very few of its subunits were equipped to face. During this short war, the BRN was forced to patch together an Airborne Early Warning system, use its area defense ships as pickets, and use decoy ships to lure missiles and aircraft away from its vital carriers. These adaptations were ultimately successful, but costly. If the Argentinian prosecution of the war had been more skillful and patient, this war might have turned out differently.

The final case study deals with adaptation failure. For this purpose, the study returns to the Pacific in World War II. In this case, the USN posed a challenge to the IJN with its submarine war on commerce. Japan, almost inexplicably given that it is an island nation, completely neglected the notion of commerce protection before the war, both in terms of planning and ship production. When the USN reversed its pre-war policy not to attack merchant shipping, however, the IJN did not react with any urgency. This was in part due to the limited effectiveness of the U.S. campaign, but also because the Japanese anti-submarine warfare effort was divided and under-resourced. Organizationally, the IJN faced many of the same problems as the BRN in World War I. While the IJN possessed the resources to combat the submarine problem, it ignored it, even as its severity increased. By the time the IJN recognized that the submarine problem needed to be addressed, it had lost the ability to respond effectively. Furthermore, the IJN failed to use the weapons and personnel it did have effectively.

The conclusion of this book summarizes the findings from the case studies and explores the implications of those findings. These implications, in turn, have important ramifications for both policy and further research, which I explore in further detail. In short, innovation is not a panacea. Emphasizing innovation in defense planning at the expense of organizational flexibility can have significant negative consequences. Outcomes in war depend upon the actions of at least two actors, so one's opponent is almost guaranteed to do something unexpected. The only way to deal with the unexpected is to be flexible.[6]

1

Setting Up the Puzzle

VIEWED IN THE larger context of the causes of victory, innovation's effect is hard to isolate. Victory is often "overdetermined." While innovations in technology, tactics, operations, or strategy have telling effects on the battlefield, other factors such as numbers, strategic surprise, leadership, strategy, and geography invariably also play significant roles. In trying to understand just how complex it can be to sort out which factors led to victory in a particular battle or war, it helps to look at the sporting event–battle analogy. Sporting events are often compared to contests on the battlefield (especially football in the United States), yet despite the obviously benign nature of sporting events relative to warfare, there are some important differences between them. Sports have rules and officials. The size of the playing field, the number of players, the length of the game, and what the players can do on the field are all fixed parameters. The effect of these parameters is twofold. First, it makes the opponents' skill the most important factor in determining the victor. Second, it makes analysis of why one side emerges victorious relatively straightforward: ideally, whichever side plays better, wins.

The key argument in this book is that the effectiveness of innovation in war is largely determined by the opponent's ability to organize against it. In other words, an innovation's effectiveness is a function of the adversary's ability to adopt it, counter it, or both. If one carries the above sporting analogy to the battlefield, the opponents not only try to fight with skill, but also try to change the number of players on both teams, affect the size and shape of the field, and change the length of the contest for their own advantage. What would be considered grievous acts of cheating in the realm of sports are in fact strategies or "innovations" of various sorts on the battlefield. Innovations from one side spawn countermeasures from the other. This makes determining innovation's effect on victory in wartime exceedingly difficult. Hence, this book has focused on adaptation to innovation. While this more detailed focus does not completely eliminate the complexity problem, it provides a better understanding of innovation's effect by measuring opponents' ability to counter a particular innovation.

In this chapter, I will set the theoretical foundation for this line of inquiry. I will define military innovation and demonstrate that there is a significant opportunity for original research in this subject area. Next, I will discuss the opponent's ability to organize to meet the new threat of an innovation. I will use a combination of political science, history, and business literature to develop this independent variable. Finally, I will discuss methodology. Given the nuanced process of innovation and adaptation, the case study approach is most desirable; the subsequent chapters that make up the rest of the book are the case studies. I will examine the following: the British reaction to the U-boat menace of World War I; the U.S. reaction to Japanese innovations in naval aviation in World War II; the U.S. reaction to North Vietnam's air defenses during the Vietnam War; the British reaction to the Argentinian sea denial operations during the Falklands War; and the lack of effective Japanese adaptation to the unrestricted U.S. submarine campaign in World War II. In the conclusion, the results of the case studies are discussed, along with their implications.

Given that the battlefield effects of innovation can only be understood in the context of the relationship between innovator and adapter, decisions to innovate should consider the organization and doctrine of likely adversaries. The most effective innovations pose complicated challenges that transcend organizational boundaries in their impact. To hedge against battlefield surprise inflicted by innovation, military organizations should emphasize flexibility, agility, and creativity in crises. Militaries must be able to anticipate primary, secondary, and even tertiary effects of a potential adversary's new technology and doctrine on their own capabilities and organization. Furthermore, they must be prepared to adjust their organizational arrangement after conflict begins in order to better adapt to innovation.

Military Innovation

Problems, sources, and issues of military innovation have spawned a large body of literature.[1] My definition of military innovation is: a new mode of operation of a military service's combat arm that yields significant or decisive advantages, at least initially, to the innovator. The innovation may be developed in peacetime, but due to the timing of its use (i.e., immediately prior to armed conflict) or other factors, the adversary's reaction to it comes during wartime. It is recognized that not all innovators get a sizeable "leg up" when they introduce their new modes of operation. Cases in which they do acquire an advantage are the most analytically useful, however, because these arguably require the target of the innovation either to react in some way or to suffer from the disadvantage. One way to understand achievement of military objectives is the imposition of one's will upon the adversary. If a military operation (and by extension, an innovative method of operation) fails to impose a change in behavior on an adversary, then it was not a very significant innovation in the first place. Therefore, I do not examine "unsuccessful" innovations.[2]

Concerning land warfare, an innovation may mean a change in how the infantry, armor, or artillery arms fight on their own or in concert with one another. The same is true with surface, subsurface, and air subunits in naval warfare. In air warfare, an innovation involves the operational behavior of strategic bombers, air superiority, and tactical aircraft. Because this study aims at evaluating the effect of innovation by studying adaptations to it, "new" is a relative term in this context. A new mode of operation is one that is out of the realm of organizational experience or memory of an opponent.

My definition of military innovation draws heavily from that of Stephen Rosen. Military innovation, he argues, involves the change of methods of operation of a combat arm, a change in the ideas governing how that combat arm contributes to victory, the creation of a new combat arm, or a change in the relationship between combat arms.[3] My book, however, does not consider peacetime innovation outside of the context of an imminent armed conflict. It also combines the wartime and technological types of innovation. Matthew Evangelista's and Owen Coté's definitions of military innovation justify this approach. In his analysis of U.S. and Soviet development of tactical nuclear weapons during the Cold War, Evangelista takes a technology-focused approach. Technological innovations involve weapons "that portend major organizational changes, reallocation of resources, possible diminished organizational autonomy, [or] significant changes in strategy."[4] The causal arrow in his analysis runs from new technology to new doctrine, demonstrating that in a combat arena that is "machine-centric" (e.g., nuclear weapons), technology plays a major role in changing, not just improving, modes of operation. Coté defines innovative doctrine as that which "exploits radically new weapons and/or beliefs about how to fight, and destroys old, outdated doctrine."[5] Coté's perspective also demonstrates the importance of technology in operational innovation. While most important military innovations have involved far beyond simply the application of new gadgetry, new technology cannot be divorced completely from an analysis of the way a military conducts operations.

Innovation and War Outcomes

Numerous works address military innovation's significance in war, but none attempt a systematic analysis of its effect on outcomes in particular armed struggles. As I have discussed above, innovation's decisiveness is extremely hard to isolate in such a complicated human endeavor as armed conflict. Also, the vast majority of the literature focuses on why organizations innovate or not, and the various methods by which they accomplish or fail at innovation. How much and for how long these innovations exert influence on the adversary is underexamined and therefore underappreciated.

Barry Posen places a considerable amount of importance on military organizations' ability to innovate. In his study of German, French, and British doctrines in the interwar period, Posen argues that innovation (or its converse, stagnation) plays an important role in victory and defeat. For example, French military doctrine suffered from military conservatism and political neglect, resulting in an army that was ill-prepared for the fluid maneuver battles of 1940. The army favored firepower over maneuver, first because of its experience during World War I, then because the level of training among its rank and file suffered with shortened terms of service. Fighting maneuver or encounter battles requires a high degree of professionalism, something short-term conscription could not produce. Tanks were subordinated to the infantry, and there were no army units capable of supporting a breakthrough of enemy lines.

French politicians, however, sought to "buckpass" the task of balancing against Germany to its Eastern European allies and to Great Britain; hence they did not press their army to find innovative ways to combat Germany's rising military power.[6] Conversely, British politicians were dissatisfied with the Royal Air Force's focus on offensive bombing and pressed for better defensive measures. This led to the organization of Fighter Command and the development of radar, which were both instrumental in the Battle of Britain.[7] German army doctrine remained offensive in nature, even after World War I. The *Wehrmacht* institutionalized "*Hutier*" or infiltration tactics, as well as single- and double-envelopment operations. German doctrine, however, stayed integrated with Hitler's geopolitical strategy of unlimited expansion by attempting to maintain a high operational tempo once hostilities started. Massed armored formations, motor transport, close air support, and an emphasis on deep penetration and disruption of enemy rear areas were all intended to induce rapid collapse of enemy resistance. The doctrine remained consistent with that of World War I so, as such, was not strictly innovative, but was innovative enough to remain "integrated" with Hitler's political strategy.[8] Likewise, Rosen does not focus on innovation and war outcomes. One of his primary goals is to correctly identify the source of military innovations.

Unlike Posen, he argues that the source of military innovation comes from within a military service, not from without.[9]

Richard Overy also asserts that innovation was an important reason for the Allied victory in World War II. During the war at sea, the Allies maintained a numerical advantage in the Atlantic and built one in the Pacific. While it is tempting to argue that preponderance of resources was the key to victory, Overy argues that triumphs during the Battles of the Atlantic and Midway came when that preponderance was not yet established: "Tactical and technical innovation won the war at sea before sheer numbers came to matter."[10] Likewise, the Soviets achieved their victories at Kursk and Stalingrad through better central planning, improvement of technology and tactics, and superior morale.[11]

Kier Lieber demonstrates that technology is not the arbiter of victory and defeat. Lieber's work comes the closest to observing innovation's relationship with military outcomes. The goal of his book, however, is to challenge the "technological core" of offense-defense theory. The broad version of offense-defense theory includes such factors as geography, resources, nationalism, regime popularity, alliance behavior, force size, and military doctrine. Lieber argues that this version is "atheoretical and unfalsifiable."[12] The technological core states that "changes in military technology shift the relative advantage of attack and defense."[13] As such, his work addresses primarily technology-driven innovation. By closely examining the wartime use of railroads and the revolutions in firepower, armored vehicles, and nuclear weapons, Lieber tests the most theoretically rigorous version of offense-defense theory. He finds it wanting in that "changes in the offense-defense balance of technology have little influence on combat outcomes."[14]

Similarly, George Raudzens argues convincingly against technological determinism in war outcomes. Through a discussion of a fairly broad swath of military history, he demonstrates that technology was rarely, if ever, decisive by itself. Other factors, notably superior doctrine, training, and discipline account for victory more often than not. This was true in great power wars, wars between greater and lesser powers, and imperial or colonial wars.[15] Kenneth Waltz's neorealist theory does not address military innovation specifically, but points to the homogenization of state behavior and imitation of the great powers' weaponry as a major factor in international politics.[16] George William McNeill's landmark work analyzing the interaction of society and military force confirms this. Important innovations, such as systematic drill, improved use of artillery, fortified encampments, and regularized marching, spread fairly quickly throughout Europe during the Thirty Years' War. They led only to "local and temporary" disturbances in the balance of power.[17]

While these works shed some light on the significance of innovation and war, none systematically tests the conventional wisdom that innovation is decisive in war. It is clear that successful innovation has important effects. Indeed, Posen demonstrates that its presence or absence among the Allied militaries prior to 1940 was significant. Integration of a nation-state's political goals and military doctrine was more decisive, however. Furthermore, it is clear from the historical literature that innovation in itself is not sufficient for victory. On the other hand, if innovation does not win wars on its own but is clearly important, how important is it? What is needed in the literature is a theory that can tell us how important innovation is in any given conflict. The current literature is replete with new theories as to how and why military organizations innovate, but like the foundational works of Rosen and Posen, they do not attempt to systematically assess innovation's independent effect in war.[18] Policymakers rarely deal in absolutes such as whether to completely change their organizations or not to innovate at all; rather, they must decide on a course somewhere between those extremes.

The policy implications of this study are also important. The technological, operational, and organizational innovations surrounding the "revolution in military affairs" (RMA) have led to an intense debate over how to allocate scarce national defense resources: "Proponents of this information technology RMA (IT-RMA) argue that the US must embrace emerging technologies and rapidly transform its armed forces to guarantee its military superiority for the foreseeable future. The current technological lead, if preserved, would increase our military strength while cutting costs . . . and reducing the risk to US troops."[19] There are many consequences of aggressively pursuing "transformation" along the lines of the IT-RMA, including costs to readiness, translating military advantage into political influence, and decreased interoperability with less advanced but increasingly vital allied militaries.[20] If the potential impact of military innovation depends a great deal on the capabilities of an adversary to react to it, as my hypothesis asserts, then achieving a significant qualitative lead in one area may not mean a great deal if the adversary adapts in an asymmetric way. Also, the United States has fought two recent major ground wars with a philosophy heavily influenced by the RMA, with varied results. The initial invasions of Afghanistan and Iraq proved to be quite decisive. New methods of fighting yielded little success against the remnants of both the Afghan Taliban and the Iraqi insurgency, however.[21] These difficulties were the subject of further rounds of innovations and adaptations.

It can now safely be asserted that the U.S. "unipolar moment" is long over.[22] With the rise of the People's Republic of China and a resurgent Russia, the United States will be faced with numerous potential innovative challenges to its military capabilities. This study can potentially shed light on whether or not the costs of innovation are worth the risks. Also, as the era of "Great Power competition" solidifies and likely adversaries reveal themselves, this theory of innovation's effectiveness can point toward fruitful avenues of innovation and potential challenges to adaptation to emerging technologies and methods of warfare.[23] I now turn to developing a theory that will illuminate innovation's importance by examining adaptation.

Ability to Organize to Meet the Threat

This book attempts to explain innovation's effect on war outcomes. Therefore, rather than merely including adaptation as a lesser form of innovation, I use it to measure the effect of innovation. Adaptation is responding to a significant change in the innovator's mode of operation. Because adaptation requires creative problem-solving, many of the requirements for adaptation are the same as those for innovation. James Wilson defines an innovation as the "performance of new tasks or a significant alteration in the way in which existing tasks are performed."[24] Williamson Murray notes that "similar factors drive successful innovation in peacetime as drive successful adaptation in war."[25] Successful adaptation requires a "significant alteration" in the way in which a military organization accomplishes its mission. It requires an organization to have an internal arrangement that enables it to process and utilize information effectively, as well as sufficient resources (both human and capital). Adaptation is as demanding, disruptive, and complicated as innovation.[26]

The degree of success in the process of adaptation has a decisive effect on the success of an innovation. Organizational structure is at the root of my analysis as the independent variable. Timely and successful adaptation requires effective organizational structure, and it manifests itself in the ability to form a "dedicated organizational subunit" (DOS) or "top-level review panel" (TLRP) designed to tackle the problem posed by the innovation. I will define the key elements of this ability, drawing from the political science, military history, and business literatures. The discussion of resource availability can rarely be divorced from military matters, and I discuss resources in-depth below and in the cases. Resources by themselves, however, are not decisive; often an organizational structure affects resource allocation more than resource availability.

Top-Down Organizational Response

Effective organization is the hallmark of successful adaptation. Organizational structure affects every step of solving the adaptation problem: recognition, analysis, creation of a solution, and implementation of that solution, including allocation of resources. But what exactly is effective organization? The presence of or the ability to organize a DOS or a TLRP is necessary for timely and efficient adaptation. These are purpose-built entities that represent the larger organization's response to the problem.

While many, if not most, innovations have wide-ranging effects throughout adapting militaries, they are often created with a particular class of targets in mind. In effect, they impose the requirement to perform new functions on a portion or portions of the adapting organization—its subunits. What are subunits, exactly? Subunits are components of the military organization that have distinct missions, tasks, and, in some cases, subcultures. The type of innovation determines which subunits are relevant. The first task of my case studies is to reveal the context, or bureaucratic landscape, that results from both the internal arrangement of the adapting organization and innovation's imposition of new tasks. This means that what constitutes a "subunit" varies from case to case. It could be an element on a general staff, a bureau, a set of operational units on the front lines, or a hastily convened committee. As Wilson notes, organizations are more than just boxes and lines on a chart.[27] Operational units, for example, are supported by a host of tailor-made administrative, training, and logistics units. Thus, "battleships," "carrier aviation," and "Wild Weasels" include not just the ships and aircraft themselves, but also an organization's systems of training, logistical support, and doctrine development.

There are four possible organizational responses to meet the innovation. First, the organization could have no subunits with the responsibility to address the innovation. This will make adaptation hard for the organization, as no one subunit will consider it as part of its core interests, and will most likely leave it for some other entity to accomplish. Second, there could be multiple subunits that can claim responsibility for adaptation. This, in turn, can have two results. On one hand, the adaptation task may be such a peripheral concern for all of those subunits that none of them wish to divert time and resources to something they consider outside their core responsibilities. With no entity willing to take on the task of adaptation, the most likely outcome is adaptation failure. On the other hand, multiple subunits could see the adaptation task as a way of gaining influence and compete with one another for the responsibility. In this situation, time and resources better used for the adaptation itself are wasted in this competition between subunits, a situation that is also likely to result in adaptation failure. Third, the organization could form a TLRP, a temporary body convened by its leadership to study the problem and develop a response. The TLRP then promulgates the response to the relevant subunits of the organization.[28] Finally, one subunit may have primary responsibility for adaptation. In other words, the adapting organization forms a DOS to take charge of the process. The DOS may already be pre-existing, or it may be created as a result of the innovation. Both have the same effect, although the pre-existing DOS allows for a quicker adaptation. This is the ideal organizational condition for adaptation, as the DOS can concentrate on the task at hand and coordinate the efforts of the larger organization in ways that are difficult or impossible under the first two conditions described above. Under this fourth condition, the function required by the innovation and organizational mission are the same, as the innovation is the DOS's *raison d'être*.

My theory posits that the arrangement or rearrangement of an organization's subunits has decisive effects on its ability to solve the adaptation problem in three fundamental ways. First, it shifts the interests of bureaucratic actors to better align with solving the problem at hand. Extant organizational arrangements encourage viewing and analyzing challenges in terms of existing methods of war or doctrine. This does not serve the organization well in the face of an innovative adversary. Once personnel are situated in the new subunit with its dedicated function, their interest is in serving that new function rather than perpetuating the existence of their old subunits. Second, the new arrangement sets new frames of reference for analysis of the adaptation problem. In shedding their interest in the old way of war or the previous doctrine, members of the new subunit would be more likely to change their preconceptions and biases about the causes and solutions to the problem. It is unrealistic to think that a shift in organization will eliminate biases; it is more reasonable, however, to posit that the new organization will encourage actors to be "properly biased," i.e., biased in a more productive way in order to produce effective adaptation. Third, the new arrangement facilitates problem-solving by placing needed personnel in close proximity to one another. Having relevant expertise "in the room" should speed up the process of organizational learning. This factor should not be underestimated while analyzing cases that occurred before contemporary modes of communications existed.

All organizations have missions that they aim to accomplish. To a very large extent, organizations are able to define their missions with a large degree of freedom from outside influence.[29] On many occasions, organizations seek to add missions to their repertoire, as this is a means to guarantee their continued existence. Adding missions also increases their influence within their parent government.[30] They also avoid missions that they view as outside of their core responsibilities: "In short, an organization will accept new functions only if it believes that to refuse to do so would be to jeopardize its position with senior officials or if it believes that the new function will bring in more funds and give the organization greatest scope to pursue its 'own' activities."[31] These missions are not necessarily the same as functions that need to be performed.

> Since the state normally is incapable of maintaining forces to meet all or most possible threats, the military man is usually required to establish a ladder of military priorities. Theoretically he should do this in terms of the objective requirements of military security. In reality, of course, he tends to stress those military needs and forces, with which he is particularly familiar. To the extent that he acts in this manner he becomes a spokesman for a particular service or branch interest rather than for the military view point as a whole.[32]

For example, the U.S. Army has long had the *need* for tactical air support. The U.S. Air Force (USAF) jealously guarded this mission from Army encroachment during the 1950s and 1960s. It did not consider tactical air support its core *mission*, however; rather, it placed much more importance on strategic deterrence. Therefore, the USAF's ability to perform this mission atrophied during the years prior to the Vietnam War.[33] Therefore, "mission" and a needed "function" are not necessarily synonymous.[34]

Furthermore, an innovation imposes a new social function upon a military organization. This new social function may already fit squarely within a subunit's existing responsibilities or mission, or the function may lie outside them. In other words, a subunit deals with a range of issues, and a function may fall inside or outside of that range. Subunits and bureaus do not exist in isolation; more often than not there are multiple entities all struggling to gain or retain influence and resources. Therefore, it stands to reason that multiple bureaus may claim a given function simultaneously. For example, both the Navy and the Air Force occupy portions of the U.S. strategic deterrence function, because both services possess strategic nuclear weapons and the means to deliver them.[35] When this situation occurs, bureaus will struggle over the function in question, leading them to consume time and resources for no "socially useful" purpose.[36]

An example from the bureau system of the U.S. Navy (USN) is illustrative of the problems inherent to a function that is claimed by multiple subunits. Until 1940, two bureaus held responsibility for the Navy's ship design. The Bureau of Construction and Repair oversaw the design of the hull, while the Bureau of Engineering oversaw that of the propulsion plant. This arrangement had its origins in the Navy's transition from sail to steam propulsion in the mid-nineteenth century. Ships were designed for wind propulsion, and new steam technology was grafted on to these designs. With the increasing complexity of ship design and construction, however, the distinction between "hull" and "machinery" was becoming meaningless. Steam propulsion was not just an add-on anymore; the design of the hull required consideration of propulsion, and vice versa.[37] Furthermore, the personnel of the two bureaus consisted of two different sub-cultures within the Navy. The constructors were top Naval Academy graduates who were trained from the beginning of their careers to be naval architects, and they did not go on sea duty. The engineers, on the other hand, served at sea but did not have the possibility of attaining command.[38] Construction and Repair had long taken the leading role in ship construction, but Engineering was functionally just as involved in the process by the interwar period.[39]

Matters came to a head after two controversies unmasked the inherent weaknesses in the system. These were the destroyer stability problem and the high temperature and pressure steam question. The former involved vessels of the *Mahan* class. It came to light during tests that these ships would be dangerously unstable in the "light condition," meaning that when the ship was low on fuel, its distribution of weight would mean that it was liable to capsize because the ship was top-heavy. Construction and Repair had resisted assertions from the contracted builder that this would be a problem. Additionally, "following the original tests the authorities held no debate as to the sufficiency of reserve stability in this class of destroyers and the major issue thereafter became the placing of responsibility for the undesirable characteristics."[40] In short, the bureau system had produced an entire class of vessels that would have trouble staying afloat under certain circumstances. Once this major problem came to light, the two bureaus spent considerable effort not in solving the problem but in blaming each other for it.

Concerning the use of high pressure and temperature steam in ship propulsion, the Bureau of Engineering wanted to take advantage of new engine technology in order to increase power and efficiency. This would enable ships to get better performance out of their steam plants. Construction and Repair was opposed to exploring this avenue of ship design. It became clear, however, that the debate was based more upon the bureaucratic loyalties of the actors involved rather than the merits of their positions.

The Assistant Secretary of the Navy stated that these two controversies were a "glaring example of the lack of coordination between the bureaus . . . and also an example of the need for a change in organization and procedure which would preclude the possibility of a re-occurrence of a condition of this kind."[41] The public became aware of these controversies, and the obvious wastefulness of building a ship that was inherently unstable prompted congressional inquiries. That and the demands of the naval rearmament program of 1938 prompted the establishment of the Bureau of Ships in June 1940. The two older bureaus were disestablished, and their responsibilities were taken over by the new one.[42] Because of the old bureau arrangement that divided one task between two different (and often adversarial) subunits, however, the Navy had been suffering inefficiencies in its ship design and construction for almost eight decades.

Once we have determined the new function imposed on the adapting organization by the innovation, we can then determine which subunit of the adapter will be primarily responsible for responding to it. Because each military organization arranges its subunits differently, and because the innovation may not neatly correspond with any particular subunit's area of responsibility, this is not as straightforward as it might first appear. If the new function encapsulates one subunit, effective adaptation will be easier than if the function overlaps more than one subunit or lies outside of any subunits' existing realms of responsibilities, other factors being equal. A new function that does not align neatly with one subunit presents a host of potential difficulties to the adapting organization. If two or more subunits have a legitimate claim to deal with the adversary's innovation, what I term a "competitive adaptation requirement," the subunits' desire to expand their missions and resources could encourage bureaucratic infighting. Conversely, an innovation that lies outside of any subunit's area of responsibility may lead all subunits to shun additional output of scarce resources toward adaptation because the new function lies outside of their core mission areas. I term this condition a "shunned adaptation requirement." The relationship between the new required function and arrangement of the adapting organization matters because, as Anthony Downs states, "*the only way to create undiluted advocacy for any given set of functions is to create a new bureau solely responsible for discharging them.*"[43]

Both competitive and shunned adaptation requirements represent "diluted advocacy" within an organization. If a military organization already has a subunit that is ready and willing to take on the challenge of adaptation, this is functionally equivalent to creating a new bureau.

The relationship between subunits and potential types of adaptation function requirements is best envisioned as the requirement graphically superimposed on the adapting military's organizational chart (see figure 1.1).

Some historical cases of the effect of the presence (or lack) of dedicated subunits are helpful in illustrating the concept. These cases provide important clues about the significance of superior organization in efforts to innovate and adapt among militaries. They also justify an organizational approach to understanding innovation and adaptation.

Allan Millett's amphibious warfare case illustrates the effect of a shunned adaptation requirement. He asserts that successful adaptation "must have a foundation in intellectual commitment, and a major embrace of a new mission," not just manpower, publications, and professional education.[44] Neither Japan's army nor its navy backed up their amphibious capability with this type of commitment. Japan divided its amphibious forces between the army and the navy, and those services viewed amphibious operations as peripheral to their central missions. Also, once the initial campaigns of conquest in the Pacific were complete, and Japan felt that it had an adequate strategic buffer around the Home Islands, its strategy turned toward defense. Amphibious operations

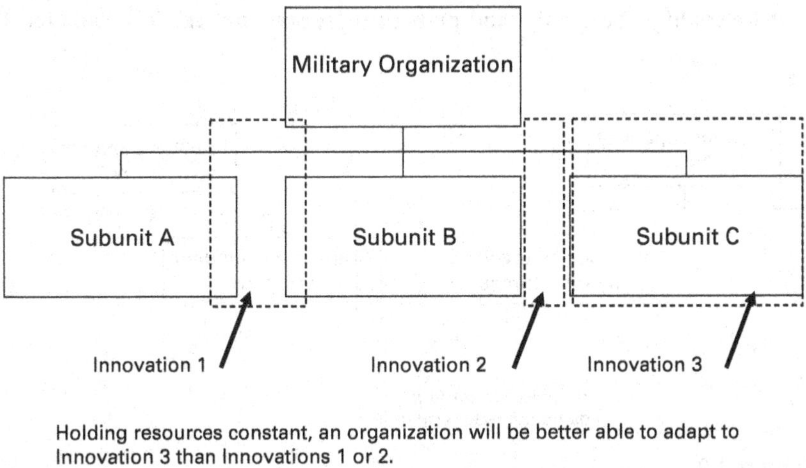

Figure 1.1
The adaptation requirements of an innovation and the relative ability to adapt.

were inconsistent with this new orientation, and both services allowed the mission capability to atrophy. Figure 1.2 illustrates this relationship.

Great Britain was relatively advanced in theorizing about amphibious operations, but its "thinkers" on the subject had no "firm organizational foundation or funding." In stark contrast to these other two powers, the United States had an independent Marine Corps that committed itself to amphibious warfare.[45]

David Johnson's examination of the U.S. Army from 1917 to 1945 provides more evidence that organizational characteristics impact how effectively militaries adapt to their environments. In fact, "internal arrangements" account for the lack of innovation by the U.S. Army before World War II.[46] The armor revolution in land warfare was a competitive adaptation requirement for the U.S. Army. Both its infantry and cavalry arms gained partial control of deciding what the role of tanks would be. Hence, U.S. tanks were designed for infantry support or traditional cavalry duties (i.e., scouting and exploitation), and were ill-suited to facing German armor designed to fight other tanks head-on. The U.S. Army would have been better served by an armor subunit dedicated to tank warfare. While Johnson labels these failures as those of innovation, much of the problem can be described as failure to adapt. Innovations in armored warfare and close air support had already been accomplished by the Germans, with telling effect. After the U.S. Army failed to innovate, its internal arrangements inhibited it from meaningful adaptation until the costly land battles in Europe, when it developed better air-ground coordination. Again, Johnson describes this as "bottom-up innovation," but we could just as easily (and perhaps more soundly) call it "adaptation,"

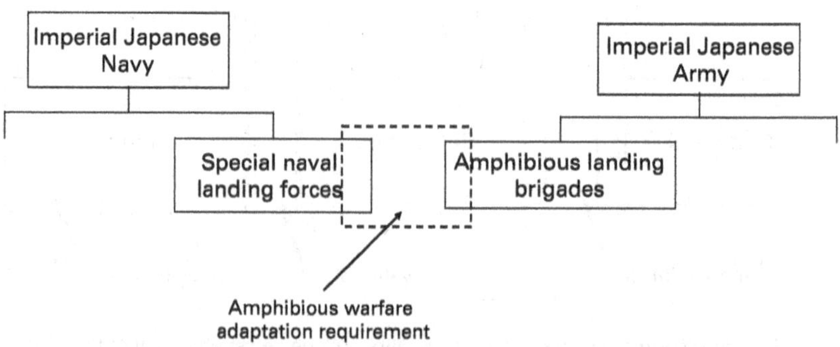

Figure 1.2
Japanese amphibious warfare subunits relative to the amphibious warfare function requirement.

because the Army was playing catch-up with the *Wehrmacht* after its belated entry into World War II.[47] Figure 1.3 illustrates the U.S. Army's organization relative to the adaptation function requirement.

In her discussion of the diffusion of carrier air power, Emily Goldman states that strategic necessity is the incentive for adoption of a technology or idea, but capacity determines how states react and what innovations they pursue.[48] She argues that capacity is determined by available resources, bureaucratic rivalry, administrative authority, and compatibility between the innovation and the organization. Great Britain was a leading innovator in carrier aviation, but because of the Royal Navy's defensive orientation (focused on protecting imperial lines of communication) and lack of control over its air arm (controlled by the Royal Air Force during 1918–1940), it ended up with offensively weak carriers (see figure 1.4). The result was that British offensive carrier air power atrophied. The British Royal Navy lost control over its striking arm, and then the Royal Air Force ignored it.[49]

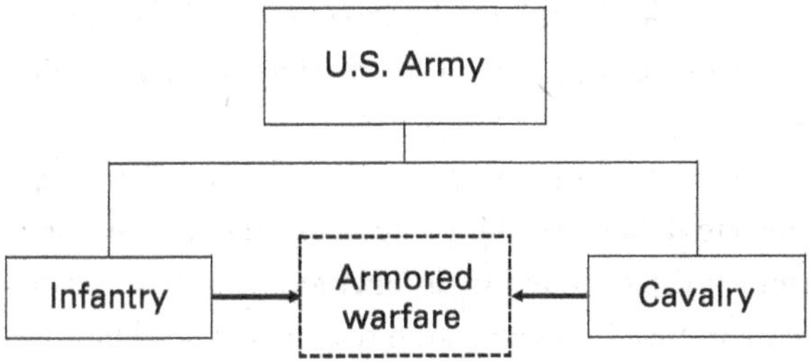

Figure 1.3
The U.S. Army's organization relative to the adaptation requirement.

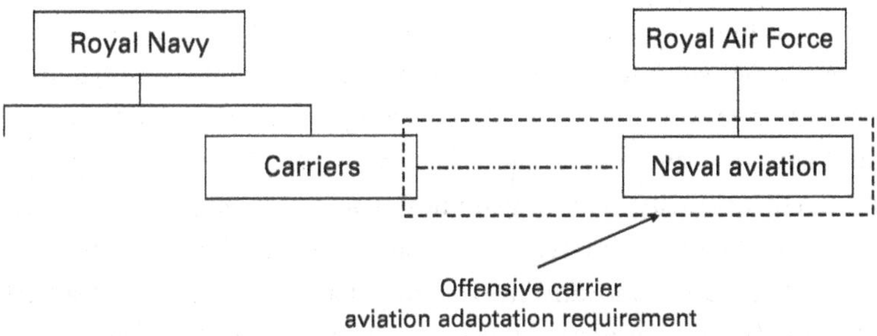

Figure 1.4
The organizational arrangement of the British Royal Navy Air Arm relative to offensive carrier aviation development, interwar period.

The degree to which an organization is centralized also affects its arrangement relative to the innovation. On one hand, too much rigidity in an organization can stifle effective adaptation to nasty surprises it encounters on the battlefield. On the other hand, however, an organization that is too loosely ordered will suffer from difficulty in disseminating information and implementing adaptations throughout its component parts. A military organization must strike an effective balance between the twin imperatives of creativity (aided by decentralization) and implementation (aided by centralization). The works of Harvey Sapolsky and Matthew Evangelista, among others, support this argument and are discussed below.

Sapolsky's study of successful bureaucratic innovation by the USN's Fleet Ballistic Missile (FBM) program points to numerous reasons for that success. Besides taking advantage of the latest managerial technologies and laying claim to its own niche within a broader U.S. nuclear strategy, the organization developed the ability to manage technical complexity by decentralizing much of its research effort, encouraging competition among its subunits, and striking a balance between flexibility and discipline in its operations.[50] The FBM program was able to successfully integrate the numerous complicated systems required for an operational Submarine-Launched Ballistic Missile (SLBM) force.[51]

While the competition between the components of the FBM program was instrumental to its success, this comes with an important caveat: all parts of the program were committed to an operational SLBM system. Sapolsky's study demonstrates that an organization that creates a subunit, which allows a group of people to 1) think about a problem, and 2) focus their efforts to produce a solution, will more likely be successful in innovation than an organization that does not do these things. While the FBM program was itself decentralized, it represented an important level of centralization: the USN possessed a unitary, coherent subunit with one goal, the deployment of a complete SLBM weapon system. This illustrates the fact that an organization faced with its enemy's innovation must develop an effective counter, or suffer accordingly. This requires a degree of innovativeness in the adapting organization, so that capacity to innovate indicates at least some capacity to adapt.

Evangelista also comments on characteristics of organizations that affect organizational innovativeness, stating that centralization of leadership inhibits incentives to innovate, but more effectively enables mobilization of resources behind an innovation that is adopted by the organization.[52] He further distinguishes between formal and informal lines of communication within organizations. Formalization is the degree "to which an organization imposes set rules and procedures for its members to follow," and greater amounts of it generally impede innovation but encourage implementation, as does centralization.[53] High interconnectedness, or the level of personal networks within an organization, encourages the dissemination of ideas and encourages innovation.[54]

The business administration literature provides some useful theory for discussing optimal organizational arrangements. Tom Burns and G.M. Stalker identify systems of management that represent "two polar extremities of the forms which such systems can take when they are adapted to a specific rate of technical and commercial change."[55] In a relatively stable (unchanging) environment, "mechanistic arrangements" of organization are suitable. These arrangements have standardized procedures, high specificity of roles, power centralized at the top of the organization, and standard sources of information. "Organic arrangements" have low specificity of roles, authority based on knowledge, and a flatter, more subordinate-centered power structure. Organic arrangements are more adaptable to complex (changing) environments, because they are more open to varied sources of information.[56] A third type, a "bureaucratic arrangement," is a mixture of the first two, where both position and expertise form the basis of authority within the organization.[57] The ability to reorganize for adaptation is greatest in the organic arrangement, and least in the mechanistic arrangement. It is difficult to envision a military organization that looks more organic than mechanistic. Max Weber describes military organizations as quintessentially bureaucratic, which would place them firmly in the middle of the mechanistic-organic continuum.[58] Military organizations that incorporate some organic characteristics should prove to be more adaptable than those that do not, however.

Eric Heginbotham's analysis of varied increases of military effectiveness in the British and U.S. militaries during World War II illustrates the importance of striking a balance between centralization and decentralization. Despite fighting the same opponents in North Africa and Europe, U.S. military effectiveness steadily improved, while British effectiveness stagnated. Rather than a lack of innovative thinking on the part of the British, the army's structure proved to be the decisive difference. Lessons learned in combat were rarely distributed throughout the army as a whole, inhibiting its ability to "remember" operational lessons: "In contrast, American forces benefited from a dense network of channels that allowed for effective communication within the force."[59]

Williamson Murray and Allan Millett's work highlights some important factors within military organizations that make innovation and adaptation easier or harder.[60] According to Murray, the *Wehrmacht* was a successful innovator in armored warfare due to a number of important organizational factors. First, it studied the World War I experience in depth and tolerated internal debate about it. Second, it also tolerated outspoken, contrarian, and less-senior officers such as future field marshals Heinz Guderian and Erich von Manstein. Third, there was an effective "feedback loop" within the organization, connecting thinkers on innovation with the operational functions. Fourth, there was an extensive professional military education system, which was not divided along service branch lines, through which members were able to learn from battlefield experience. This is a prime example of a rigidly hierarchical organization incorporating organic arrangements in order to foster innovation. The British and French military establishments did not share these characteristics, which contributed to their lack of effective innovation or adaptation to German methods.[61] This is no small matter, because in military organizations, "the greater competence and knowledge of the superior officer must be assumed. In operations, and even more particularly in combat, ready obedience cannot conflict with military competence: it is the essence of military competence."[62] Therefore,

> the subordinate officer must tread judiciously in pushing doctrines which seem to him to be manifestly superior to those embodied in the manuals. In particular, the subordinate must consider whether the introduction of the new technique, assuming he is successful in his struggle, will so increase military efficiency as to offset the impairment of that efficiency caused by the disruption of the chain of command. If it does, his disobedience is justified. Ultimately, professional competence must be the final criterion.[63]

Unless specifically allowed, outspoken and contrarian views can be hazardous to a military career.

The theory I have outlined above not only attempts to determine what makes military organizations successful adapters, but by extension, it also postulates on the source of wartime adaptations. Adaptation in wartime can come in various forms, but I have discovered two that appear to be most effective: the DOS and the TLRP. The source of both is within the military organization and is imposed from the top down.

This begs the question of whether there are possible alternatives to my theory about the source of wartime adaptation. As stated above, military innovation literature has a lot to say about the sources of innovation. Scholars have made compelling arguments that the origins lie with political leadership, from within the ranks of military organizations, and from competition between services and service components. Because I have established above that many of the attributes of adaptation are shared by innovation, I will convert two of them to theories of sources of military adaptation, and evaluate them as potential competitors to my internal, top-down explanation. The alternative theories about the source of adaptation are: 1) it comes from outside the organization, pushed by concerned civilian leadership; and 2) it comes from within, but as the result of competition between the organization's subunits. I will discuss each in turn.

Posen's argument is that innovative military doctrine comes from civilian leadership. Concerned with a whole range of international factors (i.e., the balance of power) that are above parochial military organizational concerns, their intervention is needed for successful adaptation. This is because military organizations are inherently conservative and reluctant to change. Military organizations oppose actions necessary for adaptation because this most often means a change in how they do business, affecting entrenched interests within them. These entrenched interests outweigh the dangers posed by changes in the balance of power. An example of this could come in the form of setting new priorities among its subunits or favoring one over the others. What one subunit gains by being able to claim the highest priority in the adaptation task, the others lose in both resources and influence within the larger organization. As "no service willingly accepts second priority, with an inferior claim on resources," subunits will resist the new way of doing business.[64] Even if it is in the best interest of the nation as a whole, organizations will be loath to accomplish change by themselves, as it may lead to the loss of resources or influence. Thus, military organizations will not adapt without close civilian supervision or intervention.[65]

Innovation also comes as a product of competition between subunits within an organization. This explanation arises out of Sapolsky's theory on the source of military innovation. His study of the USN's FBM program revealed that it was both decentralized and internally competitive. The workload was spread among many subunits, and the "jurisdictional boundaries" between them were not rigid. Therefore,

> everyone had actual or potential rivals and no one was assured a monopoly. Thus within the program everyone had a strong incentive to watch for problems in the designs offered by others while working diligently to avoid any of their own. Decentralization assured a

> manageable division of labor and competition assured the honesty of the laborers.[66]

This means that in their quest to maintain or expand their influence within the larger organization, subunits will push each other to find solutions to the problem. Each subunit would test its innovation, and some would succeed while others fail. Competition would thus mean that the best solution would rise to the fore. Andrew Bacevich presents a similar argument in his discussion of the U.S. Army's "Pentomic Division" construct. The "New Look" defense policy and budgetary pressure produced intense interservice competition between the U.S. Army, Air Force, and Navy during the Eisenhower administration. In response to the perceived dominance of the Air Force and Navy in the nuclear deterrence mission, the Army developed a new divisional structure to operate in the nuclear environment. While the Pentomic Division did not survive, Bacevich demonstrates that interservice competition was a prime driver behind organizational and doctrinal innovation.[67] As Adam Grissom notes, Owen Coté's work follows in a similar vein. In a comparison of Polaris and Trident SLBM development, Coté finds that a lack of competition between services can be detrimental to innovation. Whereas Polaris development spurred the USAF to improve its own weapons, Trident did not, in part because there was less competition between the two services in the 1980s versus the 1950s and 1960s.[68]

Because the case studies analyze adapters' organizational arrangements and abilities to reorganize, they will test my theory of top-down, internal adaptation and the two alternative theories. I term the first of these the "civilian intervention" model. I will determine if the evidence in the cases supports the notion that politicians stepped in to ensure that the military organization in question adapted properly. The second alternative is the "competition" model. That model will be evaluated as more persuasive if the evidence points to bureaucratic competition as the source of successful adaptation.[69]

Resources

Warfare is a resource-intensive business, and adapting organizations need these resources just as much as any other. Allocating scarce resources during wartime involves particularly high stakes. Common sense would imply that the more uncommitted resources in the form of materiel and personnel an organization possesses, the greater its ability to adapt should be. These assets represent resources that can be allocated to problem-solving relatively quickly, with minimal detriment to existing missions. Adequate resources take a back seat to organizational structure in successful adaptation, however. Without appropriate organizational structure, or the ability to achieve it, a would-be adapter struggles to find good solutions to problems, if it can at all.

The knowledge base of an organization's personnel is also fundamentally important, because warfare is also a technical enterprise. Knowledge can come in the form of technical expertise or experience in successfully solving previous problems. A mix of experience and expertise allows the organization to think in the new context forced upon it by the innovation.

Wesley Cohen and Daniel Levinthal identify a firm's knowledge base as significant in shaping its ability to innovate. They argue that innovation depends in large part on the ability to exploit knowledge originating outside of the firm. Exploitation of that knowledge requires some sort of internal knowledge related to that coming from the outside; otherwise the firm has little context within which to frame its efforts. They term this ability "absorptive capacity," and as this capacity goes up, so does the firm's ability to innovate and be proactive in the market.[70] While the authors focus on innovation, I submit that this knowledge base is equally important for effective adaptation, especially if the organization is to attempt sophisticated measures.

In their theories about the behavior of firms in the business world, Richard Cyert and James March identify organizational "slack" as an important asset in making innovation possible. They define slack as "the difference between payments required to maintain the organization and the resources obtained from the environment by [it]."[71] Slack provides resources for "innovations that would not be approved in the face of scarcity."[72] In other words, innovative organizations are more often than not also smart organizations.

While I do not view resources as the decisive factor in successful adaptation, their relative presence or absence warrants careful consideration in the case studies. Military history has demonstrated that superior resources in the form of population, raw material, and industrial capacity (to name a few) are decisive in the long-run, strategic context. My case studies analyze the resource considerations at play in each conflict, especially in the context of particular innovations, in order to evaluate their explanatory value relative to organizational structure.

Method

This research project is designed to isolate the independent effect of innovation through analyzing the reactions of adapters. The best method to examine the effects of innovation on military outcomes is through detailed analysis of case studies. By studying innovation in terms of its effects, and on the reaction of the innovator's target, we can ascertain whether or not the innovation process contributed substantially to an operational victory, lowered casualties, or had a negligible effect. This method is what Alexander George and Andrew Bennett refer to as "process-tracing," whereby the researcher examines detailed historical records to ascertain whether or not the "causal processes" of a theory are actually at work in a given situation.[73]

In each case, the book first establishes that an innovation in military or naval operations took place. Because the goal is to analyze the independent effect of innovation on wartime outcomes, this means that the innovation was successful, in that it had a positive effect on achieving campaign objectives of the innovator or inflicted significantly greater casualties on the adversary, and thus changed the adversary's behavior. There is good reason to focus on innovations that were "successful" in these ways. Studying an innovation that is unsuccessful has little utility in determining whether it was it worth it. At best, an unsuccessful innovation (i.e., one that neither inflicts increased casualties, nor changes adversary behavior) had no discernable effect; at worst, it cost lives and treasure to the innovator. It stands to reason that the innovation was not worth the effort in the first place. Examining the reasons that the innovator made its policy decision to pursue an unsuccessful mode of operation is a worthy, but different, project than this one. I am posing a more difficult test for innovation's independent effect on the battlefield: what does wartime experience teach us about the return on investment for demonstrably successful innovations?

What counts as successful adaptation? My research looked for behavior in the adapter that went beyond trading resources for time or otherwise simply absorbing the additional punishment inflicted by the innovation. A successful adapter is able to recover from its initial shock and pivot its main effort; at some point it perceived that method of war or doctrine "A" was not working, and it was successful in shifting to method of war or doctrine "B." This shift is then marked by reduced casualties from the innovation and improved progress of its operations. It is the job of thorough process-tracing to detect and document if and when that transition to success occurs.

All of the cases in this book have been thoroughly documented, so this first step of identifying and defining the initial innovation was a relatively straightforward task. The ensuing process of adaptation that is my analytical focus starts after the organization has recognized that it is the target of an innovation. In recognizing the innovation, intelligence would presumably play a vital role. An organization's ability to do this successfully would largely reflect the capabilities and shortcomings of its intelligence apparatus in discerning the new mode of operation correctly. Without this recognition, it should be noted, little effective or coordinated adaptation is possible. Also, the very start of each case is an examination of the bureaucratic landscape relative to the innovation: what components of the adapting organization are affected? Which of these subunits responded, which ones did not, and why? Are there multiple ways to parse the bureaucratic landscape? With this in mind, my research uncovered and examined the process by which the organizations: 1) formulated a response; 2) promulgated that response to its subunits; and 3) effectively carried out operations that incorporated the adaptations. It is through focusing on these steps in the adaptation process that I have gathered my evidence.[74] Where available, I have used primary documents, such as after-action reports, memoranda, and message traffic, which will further reveal the relationship between the adaptation and success on the battlefield. Also, primary and secondary documents should reveal how well-equipped the adapting organization was to face the innovation. When an innovator faces a sophisticated adapter with the ability to organize to meet the new threat, its rate of success should be less than when it fights an adapter with less capacity (see figure 1.5).

Again, victory in warfare is almost always overdetermined. In a long war, superior resources or geographical advantage make themselves felt, given the opportunity. Rather than using overall victory or defeat as a benchmark, I will look for wartime effects such as operational victories and significant changes in casualty rates that can be tied to the innovations and adaptations. By tracing the effects of innovation and adaptation in this way through a number

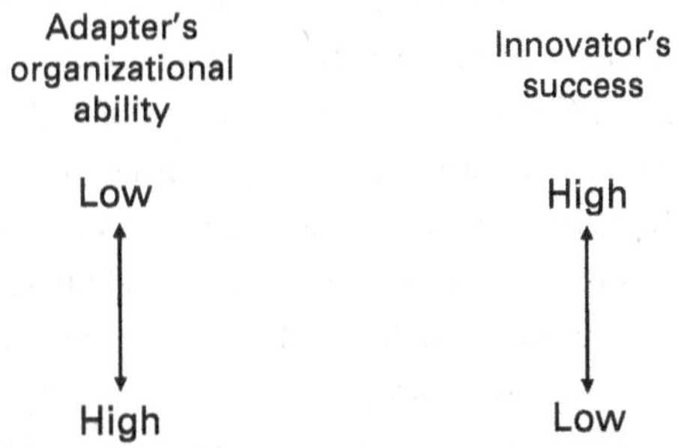

Figure 1.5
The organizational ability of the adapter vs. the success of the innovator in wartime.

of cases, we can go a long way in determining how decisive they were in the wartime environment.

Cases and Their Selection

In studying the independent effects of military innovation on wartime outcomes, I have selected cases that 1) have occurred in wartime; 2) examine an operational-level innovation; and 3) are largely maritime in character. Wartime cases are necessary to examine wartime outcomes, but are also more impactful. Operational innovations represent important changes to warfighting methods and doctrine. The operational level is also more appropriate for study because at the strategic level, much depends on geography and other factors unique to particular nation-states, which are not generalizable to military organizations across states. Finally, naval or maritime innovations represent a more robust test of the theory because of the technical and resource-heavy nature of naval warfare. These three characteristics guided my case selection, rather than an adherence to a particular kind of conflict (e.g., "Great Power war," or "counterinsurgency"), which would have unnecessarily shrunken an already shallow pool of likely cases.

I have selected cases in which, for various reasons, the adapter has not reacted to the innovation until war has begun. The innovation-adaptation relationship of course occurs outside of the wartime environment. States are constantly analyzing the capabilities of their potential opponents and reacting to their new modes of operation by land, sea, and air. There are, however, two important reasons for focusing only on the wartime innovation-adaptation relationship.

First, military organizations are most severely tested during wartime. On one hand, the degree of urgency brought on by combat operations ensures a high level of attention to adaptation by the upper echelons of the organization. As the organization suffers from the disadvantage brought on by the innovation, those in charge will both feel and exert heavy pressure to adapt. On the other hand, time constraints make adaptation more difficult. Peacetime adaptation can occur within the constraints of normal organizational routine. Because combat effectiveness and loss of life are not immediate concerns in peacetime, the sense of urgency is lower as well. The time pressures in a wartime environment far exceed those of peacetime, presumably even during an intense arms race. The combat is real, not abstract. In addition, competition for scarce resources, always a concern, is more acute during wartime. Personnel and materiel may have to be diverted from combat operations to develop an appropriate response to the innovation. The normal peacetime choice between ensuring short-term readiness and developing long-term effectiveness in a changing environment becomes a much more difficult calculation when the consequences of battle are immediate. The organizational routine is under great stress during wartime. Hence, the wartime environment allows us to more clearly see organizational dynamics at work. In terms of difficulty, wartime adaptation provides a more rigorous test of an organization's capacity to adapt.

The second reason for focusing on wartime adaptations lies in the historical importance of the innovation and adaptation combination. While during peacetime this interaction may be significant—major elements of the Cold War nuclear arms race come to mind—the high-stakes wartime environment ensures that the relationship between innovation and adaptation is significant. Given that lives and perhaps the fate of nations are at stake, the importance becomes immediately apparent.

The level of analysis on which this volume will focus is operational: the behavior of fleets, air wings, and corps. Wayne Hughes's definition of "fleet tactics" is most useful here: "operations involving coordination between multiple ships and aircraft, and the sensors that support them."[75] I term this focus "operational" rather than following Hughes's lead in order to avoid confusion with a focus on the behavior of small units or individual ships (what he terms "maneuvers," more commonly known as "tactics"). The behavior of small units at the tactical level changes with a frequency that makes analysis exceedingly difficult and of limited utility. The behavior of fleets and corps is harder to change and more durable over time. Likewise, this book will generally avoid the realm of the strategic, which involves questions of why and when armies, navies, and air forces fight, not how. Strategic-level discussions often provide important context for the case studies, however, so I will address matters of strategy when they have a significant impact on how militaries innovated and adapted.

The range of potential conflicts seems broad. Those relevant to the modern era, however, represent a much smaller sample. Furthermore, there must be sufficient historical data to examine the innovator-adapter interaction. The independent variable under examination is organization, and the dependent variable is adaptation success. The type of war is not the primary focus of this study. In order for this analysis to have maximum utility, therefore, it is necessary to test the theory as broadly as possible. This is why I have selected cases that illustrate the innovator-adaptor relationship during Great Power wars, limited wars, and those between lesser and greater powers. The majority of my cases focus on naval warfare. The Vietnam air war case provides historical depth and a more robust test of the thesis. I have selected cases from a survey of wartime innovations (see table 1.1).

I include in wartime innovations those that were developed in peacetime, but that made their first significant appearance at the outbreak of war. This is because it was necessary for the military organization to accomplish the complex task of adaptation in wartime, with all the incumbent restrictions on time and resources this entails. As noted above, the wartime environment provides a more rigorous test of an organization's ability to manage adversity brought on by an enemy's innovation.

The primary focus of the naval warfare cases is on innovations in combating capital ships and the adaptations to counter those innovations. For much of history, the capital ship has been the ship-of-the-line or battleship, followed by the aircraft carrier in the twentieth century. In order to improve the depth of this study, however, I include an example of air war as well. There are three reasons for this.

Table 1.1. Wartime Innovations

War	Innovations
Napoleonic Wars	Corps system
Wars of German Unification	General staff
World War I	Submarine warfare
World Wars I/II	Combined arms warfare (land)
World War II	Carrier warfare
World War II	Amphibious warfare
World War II	Strategic bombing/nuclear weapons
Vietnam/Yom Kippur Wars	Integrated air defense
Falkland/Middle East Wars	Anti-ship missiles/sea denial
Gulf Wars I/II	Precision-guided munitions/network-centric warfare
Yom Kippur War	Anti-tank guided weapons

Source: This table was developed from a survey of Richard Holmes, ed., *The World Atlas of Warfare: Military Innovations that Changed the Course of History* (London: Mitchell Beazley, 1988).

First, given the importance of capital ships throughout history, analyzing innovations that were at least initially successful against them will ensure that the book focuses on significant, rather than merely novel, innovations. Second, naval warfare has always imposed a high technological and expertise burden on any state that wishes to affect the balance of power at sea. Even in the era of the sailing ship, before computers, wireless communications, and the internal combustion engine, operating a single warship required a large amount of state resources. Navies have also, since the age of sail to the present day, expected their ships to have service lives of several decades; once a ship is built, it represents an incredible amount of sunk costs committed not just to that platform but also to the way of war for which it is optimized and to its support system. This also ensures that the innovations under scrutiny are significant because they are difficult. Third, because we are discussing the centerpieces of Great Power fleets, both greater and lesser powers have had an immense interest in finding weaknesses to exploit and ways to compensate for them. Maritime powers have always striven to maintain their dominance of the seas, while aspiring sea powers or even regional powers with coastlines have sought ways in which to undermine that dominance.

By extending the focus of this book to include the air war over North Vietnam, the narrative provides a greater range of innovations and historical experience with which to test the hypothesis. The Vietnam War was longer than all of the other conflicts examined in this study, but it was also limited in nature. Major wars are rare and significant, and prolonged naval wars are rarer still. Therefore, expanding the scope to include a limited-war case provides for a richer and more in-depth analysis.

The cases that I have selected for this study proceed in mostly chronological order. The first case is that of the British Royal Navy's reaction to the German use of the U-boat in World War I. Second is the study of naval aviation in the Pacific during World War II. The Japanese confronted the USN first with the improved striking power of their carriers, then with the very unusual *kamikaze* attack. Third, the Vietnam War case concerns the U.S. Air Force's adaptation to North Vietnamese air defenses. The fourth chapter discusses a short war: the Falklands War of 1982. This case analyzes the British Royal Navy's adaptation to Argentinian sea denial strategy. Finally, I consider adaptation failure, focusing on the failure of the Imperial Japanese Navy to adequately defend Japan's trade routes against U.S. submarines in World War II.

In general, primary documents have been most useful in getting an accurate picture of the adapting organization (i.e., how it was arranged internally), the resources available to it, and how the organization is positioned relative to the innovation. In other words, did the organization suffer infighting about which of its subunits would or would not bear the burden of adaptation? Did it have to compete with another organization for resources or a mandate to handle the problem? Conversely, was the organization reluctant to address the innovation in the first place, believing it to be the responsibility of another organization? These debates rarely emerge into the public sphere, but could be quite well documented in internal correspondence.

The disadvantage of this case selection is that in almost every case, the adapters were the eventual victors in the conflict. This may lead an observer to the "if it explains everything, then it explains nothing" critique. The case selection addresses this in two ways. First, the British Royal Navy in World War I is an example of an organization that first failed to adapt, then did so quite successfully. This case thus provides a very good opportunity for what George and Bennett term a "within case comparison."[76] I will analyze what factor changed in the middle of World War I that enabled the Navy to recover from its mistakes. Second, the Japanese maladaptation to the U.S. submarine war on its sea-lanes of communication provides an example of failure to adapt, or unsuccessful adaptation.

This book's operational focus produces cases in which innovators and adapters accomplished their missions with wildly varied expenditures of blood and treasure. Even if the adapter won in every case, his level of success varied in the numbers of ships sent to the bottom of the sea and lives and equipment lost. All states make a decision to go to war based in some part on whether or not they think they can win. States must, however, also consider how costly achieving victory might be. Organizational flexibility may be a key difference between efficient and pyrrhic victories.

2

The British Royal Navy and the U-boat Menace of World War I

WARS RARELY PROGRESS in the ways in which the belligerents expect. The naval dimension of World War I was no exception. Prior to the outbreak of war in 1914, the British and German Navies faced each other with colossal fleets of battleships. Both sides expected that, in conjunction with a short decisive war on land, those two fleets would meet in a single engagement that would decide who commanded the world's oceans. While that engagement did occur at Jutland in 1916, it was belated and indecisive. Although the British had turned back the German fleet for a strategic victory, the latter was not destroyed. The two battle fleets were relatively inactive for the rest of the war. Of all the advances in naval armaments that occurred before the war, the submarine eventually eclipsed those of battleship gunnery, propulsion, and armament. The British Royal Navy (BRN), which had won the pre-war naval arms race, found itself stymied by the war on its nation's shipping.

Prior to World War I and until the maturation of the aircraft carrier, the battleship was the most important weapon system deployed upon the oceans. Measured in terms of the size and quality of battleship fleets, the BRN was the dominant force upon the world's oceans. The use of submarines by Imperial Germany, however, almost trumped this balance of power. Given that a seemingly effective countermeasure, the convoy, was already in hand, how was this near-defeat of the BRN possible? In other words, why was the BRN blindsided by the submarine campaign against its merchant shipping? Why did the British maladapt? Convoy was the answer, but it took the BRN two years and more than seven million tons of shipping losses to finally adopt it. The BRN took all sorts of actions that were ineffective, wasting time and resources. Only after it restructured itself and shifted much of its operational anti-submarine warfare (ASW) focus to convoy operations was it able to reassert its naval superiority. Yet its operational focus did not entirely shift away from an offensive, submarine-hunting rubric, which left trade protection forces stretched thin even as Germany weakened.

This chapter will discuss the naval aspect of World War I, focusing on the character of the German submarine innovation. It will describe the British adaptations as the BRN gradually worked its way back to the ideal solution: the convoy. Next, the historical narrative will allow analysis of the BRN's ability to organize in terms of internal arrangement. The BRN's ability was low enough that Germany's submarine innovation nearly crippled Great Britain, despite the latter's greater naval power. This analysis runs counter to some other explanations of the BRN's lethargy in adopting convoys. I will address these counterarguments at the end of this chapter.

The Innovation: Submarines and Naval Strategy

The BRN was caught off guard by the German U-boat campaign against its merchant fleet. The adapter in this case, however, holds the somewhat counterintuitive position of innovator as well. While British Admiral Jacky Fisher rather famously provided the impetus behind the all–big gun battleship, or dreadnought, that revolutionized naval warfare in the late nineteenth century, he was also advocating the use of the submarine in war. The motorized torpedo was becoming more advanced and capable, and Fisher saw the submarine as another means of delivering it to its targets. The BRN as a whole did not accept the submarine, however. The majority of officers derided "Fisher's toys" as underhanded and "un-English."[1] Despite this prevailing attitude, the BRN gradually increased its numbers of submarines in the years prior to World War I. They would fulfill a scouting role, lurking beneath the waves ahead of the main battle fleet, alerting it to the presence of its opposite number. They would use their torpedoes to attrite the opposing battleships, helping to boost their own battle fleet's chances in the decisive engagement. The Germans possessed submarines as well, although fewer in number. They envisioned the same role for their vessels. So while the BRN failed to anticipate the upcoming U-boat war against shipping, its adversary failed to do so as well. Prior to World War I, the Germans did not plan on using the U-boat as a weapon against trade. The German Navy had rejected the necessary submarine-building program to support such an action in March, September, and again in October of 1914. Shifting resources to U-boats, in its view, would have weakened the Navy in the primary goal: to bring the British Grand Fleet to battle in a decisive action.[2]

Once war broke out, however, that decisive action was slow in coming. Both sides wanted to fight on their terms at opposite ends of the North Sea. The British wanted to engage the entire German High Seas Fleet with its Grand

Fleet because of its superiority in battleships. The Germans wished to take on only a portion of the Grand Fleet with the entirety of the High Seas Fleet in order to defeat the BRN piecemeal. The result of these strategies was no battle at all, which hurt the German Navy's morale through inactivity. Besides, the overall German war strategy demanded that the High Seas Fleet remain intact so that it could protect the Baltic coast from Russian invasion. Engaging the British fleet on its terms would endanger that role. Into this situation of forced inactivity stepped the submarine. With a small displacement and slow speed, submarines could not match the open-ocean performance of surface vessels such as battleships, cruisers, or destroyers. An early battleship-hunting cruise proved ineffective, but demonstrated that submarines could hunt surface vessels despite their limitations. When a U-boat sunk a merchant ship without German commanders' permission in October 1914, it established the offensive potential of the submarine to a navy receptive to a new strategy: "Accident, prestige, service pride, internal politics, British naval policy, and the course of the land war" all took a hand in shifting German naval strategy toward the U-boat in late 1914.[3]

German submarine warfare innovation came in two iterations. In the early years of World War I, German submarines pursued a semi-traditional *guerre de course*, otherwise known as privateering. Traditional *guerre de course* required that a merchant raider or privateer stop and inspect its target to determine its nationality and cargo. If the vessel was under the enemy's flag, or the cargo could reasonably be assumed to be destined for a belligerent nation, this justified the raider in taking it as a prize of war, the merchant's crew being taken prisoner, and the ship being sailed to a friendly port by a skeleton crew from the raider. Destroying the ship and cargo was a last resort that was rarely taken, because the raider's crew would receive no financial benefit if the prize vessel did not reach a friendly port.[4]

Throughout the era of the privateer, the world's navies (including the BRN) recognized that convoy was the most effective counter to this method of warfare.[5] By the mid-nineteenth century, international law made privateering illegal. Sea trade became an increasingly large and complicated economic activity. While it was vital for feeding the British people and supplying the nation, civilian policy-makers and naval leaders came to view the protective measures of convoy as more economically damaging than the potential harm by merchant raiders. Convoys took time to assemble, and the complications associated with that process caused delays and inefficiencies. As British trade depended upon an uninterrupted flow of goods, these delays and inefficiencies had a similar economic effect to actually losing the cargo to merchant raiders. The cargo "lost" in this fashion would often amount to more than a raider fleet would be able to claim from unescorted merchants. With these assumptions in place, the Admiralty unsurprisingly took a *laissez-fare* attitude toward merchant shipping. Government interference would do more harm than good, so better not to impose convoy on the merchant fleet. Besides, dispersal of trade throughout the wide expanse of the world's oceans was preferable to the easy target a convoy would make. This was the dominant attitude within the BRN at the outbreak of World War I.[6]

While the oceans on which British trade traveled were vast, the routes it took converged as they neared the British Isles. Dispersal of trade by itself proved to be no obstacle to German submarines. From 1914 to 1916, they simply took up positions where the trade routes converged. There were ample targets. The submarines surfaced when the ideal target, a solitary merchant ship, approached. The submarine would then force the merchant crew to abandon ship in lifeboats, and then sink the merchant with gunfire. There was no room on board a submarine for prisoners, and no spare personnel for prize crews. So while the merchants were stopped and warned before they were attacked, the submarine had to destroy the ship and its cargo and abandon the crew on the high seas. There were a few notable exceptions to this practice, in which ships were sunk without warning, but these were the exception rather than the rule during the first restricted U-boat campaign.[7]

The second iteration of the U-boat campaign dropped the last vestiges of similarity with traditional *guerre de course*. After the BRN began arming merchant ships with deck guns at the outset of war, even the nicety of warning had to be abandoned lest the submarines themselves become the targets. In 1916, the Germans first attacked where protection of merchants was weak, such as in the Mediterranean. The arming of merchants was so extensive, however, that U-boat commanders soon had to assume that all merchants were armed. In early 1917, they adopted "unrestricted submarine warfare," or sinking without warning, as a result.[8] All shipping (besides hospital ships) going to and from the British Isles was fair game. Whereas the submarine's deck gun was the primary weapon of the restricted campaigns, the torpedo became the primary weapon in the unrestricted operations. In order to avoid the deck guns of defensively armed merchant ships and decoy ships, the U-boat would fire torpedoes at its target while submerged, without warning. This obviously caused much greater loss of life among merchant sailors and civilian passengers on the high seas.

In hindsight, the potential of the U-boat as a raider is clear, yet neither navy anticipated the submarine in this role prior to the war.[9] The German

Navy only turned to *guerre de course* with submarines after the war had begun. Submarine warfare in this period fits the definition of wartime innovation, despite both countries possessing the technology before the war.

If one examines the British Admiralty's own correspondence, it becomes clear that hindsight was not necessary to foresee the German progression to unrestricted submarine warfare. First, there were the aforementioned isolated incidents in which merchants were sunk without warning by torpedoes. More significantly, in discussing the effects of current ASW policy, the Admiralty reasoned that these policies would force the U-boats to rely on torpedoes launched while submerged. The director of the Trade Division received numerous proposals for solutions to the U-boat problem from inside and outside the navy. One of these suggestions was that submarine crews that had "molest[ed] shipping" be punished severely if captured, to serve as a deterrent for others. Besides being impossible to prove which submarine had committed which offense, the director responded, this action would be harmful to merchant crews because submarine crews, in an effort to escape punishment, would cease to warn merchants of their attacks.[10] The defensive arming of merchant ships also threatened to bring about the same result, if even more directly. Instead of the abstract threat of possible future legal trouble, submarine crews faced the very immediate threat to their safety posed by a merchant's deck guns. The only sure way to avoid those guns was to remain submerged. The two means were very different, but the result was the same: submarines would eventually begin to sink merchant vessels without warning.

Given the deadliness of the torpedo and the later effectiveness of the unrestricted submarine campaign later in the war, why would the British Admiralty pursue such a course of action? The answer lies in the fact that the BRN considered the torpedo an incredibly inefficient weapon when used against merchant ships. In the restricted submarine campaigns, U-boats only needed to expend a few shells from their deck guns to sink a vessel. The BRN's reasoning was that U-boats had many more shells than torpedoes onboard, so if they were forced to use the latter, their cruises would be shorter and they would become less effective. Unrestricted submarine warfare would also have adverse consequences for Germany *vis-à-vis* neutral countries. This reasoning is found in John Jellicoe's correspondence with the Admiralty in late 1915. He asserted that a scheme of merchant traffic routes close to shore, regularly patrolled by ASW forces, would force submarines to use their torpedoes by keeping them submerged. This would be an advantage, not a disadvantage.[11] Intellectually, the BRN knew that its measures would force the Germans to eventually adopt sinking without warning. They underestimated the efficiency of the torpedo, however, while overestimating the potential of neutral anger over its use as a deterrent. In advocating decoy ships as the most effective counter to the U-boat, the Admiralty stated that

> You must send ships where the submarines operate and attract the submarines to them. If they will not attack them on the surface it is true we can do nothing, but if they adopt a policy of only attacking in submerged condition, they will do far less harm and be more likely to sink some neutrals and increase the feeling against Germany.[12]

While unrestricted submarine warfare eventually brought the United States into the war, U-boats armed with torpedoes almost succeeded in defeating Great Britain.[13] Certainly, submarines attacking when submerged did not cause "far less harm" than those on the surface. From its internal correspondence, the BRN seemed to make the calculation that inefficiency of the torpedo, together with the diplomatic fallout associated with its use, would serve as a deterrent against unrestricted submarine warfare.

The shipping losses of the British merchant fleet throughout the war reflect the *ad hoc* nature of the German strategy. Fewer than 3,000 tons of shipping were lost to submarine action in the last months of 1914, but 1915 saw a dramatic increase—from about 17,000 tons in January to a maximum monthly toll of 170,000 tons by the end of that year. After fits and starts due to diplomatic wrangling with neutrals (especially the United States) and internal disagreement about submarine employment (discussed below), the losses inflicted broke the 200,000-ton mark in September 1916. Thereafter, it did not dip below 300,000 tons a month before April 1918.[14] These heavy sustained losses reflect the German shift to unrestricted submarine warfare and the BRN's failure to adopt the convoy in a timely fashion. The jump to above 200,000 tons per month also represented a crippling escalation for the British, as their shipbuilding industry could no longer build enough ships to replace those that had been sunk.[15]

The Adaptations, Phase I: The British Royal Navy's Offensive Strategy

Once an innovative method of warfare makes its appearance on the battlefield, the adapter can pursue many different options in attempting to redress the decrease in its ability to accomplish its mission. Despite neglecting the use of mercantile convoys for two years, the BRN pivoted and made numerous subsequent attempts to deal with the submarine menace.

Immediately prior to the war, then–First Lord of the Admiralty Winston Churchill issued a memorandum on the matter of trade protection. He argued that the most important consideration, given Great Britain's dependence on seaborne trade, was that trade must be kept moving. Large numbers of merchant ships servicing British commerce across the large expanse of the world's oceans would enable the country to absorb shipping losses and continue to support itself.[16] This reflected long-standing British policy and reinforced it for the coming war. When Churchill wrote this memorandum, however, the primary threat to trade was considered to be the surface raider, not the submarine. Even though the threat shifted from surface to sub-surface, this philosophy of trade protection remained unchanged through more than two years of war, even though it was not successful. Policy did change in 1917, however; hence, the BRN pursued the countermeasures adaptation in two rounds. First, it used offensive countermeasures during 1914–1916. Second, it gradually introduced the convoy system in the spring of 1917, fully implementing it in September of that year. The first round of adaptation was a dismal failure. The second round proved to be the decisive counter to the U-boat. The narrative below explains how the BRN changed its policy of trade protection.

After four months of war, the Admiralty requested that all major commands (the Grand Fleet, Dover Patrol, various harbor commands, etc.) report their methods of fighting submarines. This memorandum and the responses to it are illustrative in two ways. First, it demonstrated that the BRN had no coherent, top-down approach to fighting submarines. In December 1914, the BRN was using many different methods of combating U-boats, but they varied from one command to another.[17] Second, the memorandum illustrates the majority of the BRN's philosophy toward the threat. Despite some merchant shipping losses, no commands suggested the use of the convoy. In general, the various parts of the BRN were using combinations of mines, gunfire (when they found surfaced U-boats), decoy ships, ramming, and nets (to entangle a submarine and indicate its presence).[18] All of the commands heartily advocated offensive measures that would harass and disrupt the U-boats in their missions before they had a chance to harass and disrupt British naval and merchant shipping. As the various departments of the Admiralty staff reviewed the fleet's methods, they endorsed the lower echelons' offensive measures.[19] These boiled down to U-boat hunting patrols and minefields designed to trap submarines in their bases.

It soon became clear that the best vessel to "harass and disrupt" U-boats was the destroyer. The Grand Fleet possessed the largest concentration of modern destroyers in the BRN, and the fleet's commander at the time, Admiral Jellicoe, jealously husbanded these valuable craft. They protected the battleships of the fleet from their counterparts in the German fleet, as well as submarines. Additionally, their torpedoes were valuable weapons in their own right during a fleet action. Yes, the destroyer was much better at hunting submarines than a trawler because of the former's speed, communications, and well-trained crew, but Jellicoe wrote: "It is . . . a most extravagant thing to use the First Fleet Flotillas to any great extent for hunting submarines. It wears the vessels out, both as regards material and personnel, with the result that, when required for working with the Fleet, they are not available."[20]

Furthermore, Jellicoe argued, there were not enough destroyers for a proper ASW "offensive."[21] If, on the other hand, there were sufficient numbers for such an offensive, the action should be carried out. Jellicoe forwarded a "representative sample" of ASW suggestions from the fleet in mid-1915. The suggestion called for "offensive concentration," in which two massive fleets of destroyers and other smaller craft would harass and hunt submarines in a given area for a given length of time. The two groups would coordinate with each other and trap U-boats between them. Jellicoe liked this idea, writing, "I am convinced that the policy of offensive concentration which is advocated in the paper enclosed is the sound method to adopt if not carried out to an extreme."[22] This further demonstrates that the prevalent attitude of the BRN up and down the chain of command was that offensive measures should be the dominant form of response to the U-boat.

Throughout history, navies have fostered a culture of aggressive action.[23] The BRN at the time of World War I was no exception, and in fact epitomized this. The effects of this aggressive spirit manifested itself throughout the organization. During the Battle of Jutland, three BRN battlecruisers blew up in quick succession, leading Admiral David Beatty of the battlecruiser force to dryly comment, "There's something wrong with our ships today." It turned out that BRN battlecruisers, which as a class were weakly armored compared to battleships, also suffered in comparison with German battlecruisers in the protection of their magazines, with catastrophic results: "Those three battlecruisers were, quite simply, a gambling loss, part of the price paid for the element of self-confident aggression in British naval tradition. The firm belief existed that there was something weak and even effete in giving too much attention to caution and protection from the enemy."[24]

When it came time to consider what to do to counter the U-boat, it was hard for the BRN to shake its commitment to offensive and aggressive action:

> In the end it came down to a state of mind in which the Nelsonian principle of attack (albeit distorted) was applied to training of officers and men and the design and construction of men o' war. In short, defense was an unacceptable principle, offense correct—right-minded, valorous, glorious, and correct. To protect merchantmen, to scurry about them like a sheepdog, was *defensive*.[25]

When the various commands went without central guidance on ASW measures, it is not surprising that they fell back on offensive measures.[26] It is also not surprising that those suggestions that were passed up the chain of command were, for the most part, offensive in nature.

The British Royal Navy's Organizational Structure

In 1914, the British Admiralty Staff consisted of four divisions: Operations, Intelligence, Mobilization, and Trade, along with the highly specialized Transport Department (see figure 2.1).

The Operations Division's core responsibility was the movement of BRN fleets and ships; Intelligence collected information necessary for war plans; Mobilization dealt with war arrangements, such as the staffing of ships; Trade,

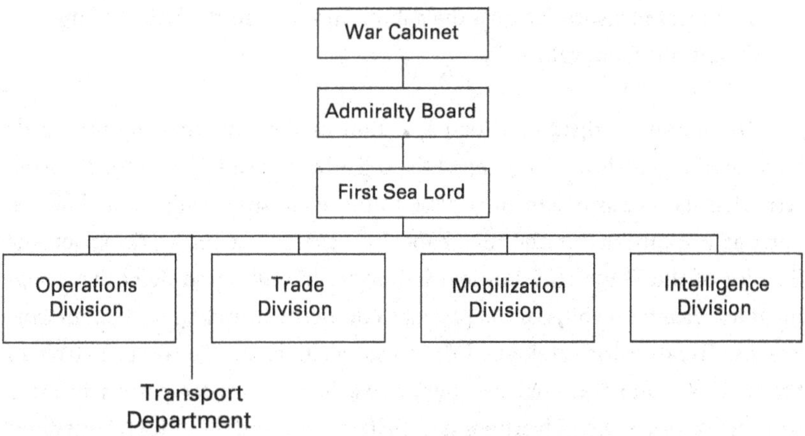

Figure 2.1
The British Royal Navy's Admiralty organization, 1914.

which had been formed from the Operations Division, concerned itself with the movement of merchant ships and their protection, as well as orders for trade routes and the diversion of merchant ships.[27] With troop transports, "certain blocks of work were dealt with by flag officers with special knowledge of it."[28] Moving large numbers of soldiers across the English Channel to the Continent became the responsibility of the Transport Department, which started work after the war began and was led by the outgoing Chief of the War Staff: "The immense demands for transport were met smoothly and punctually by the Transport Department, and the War Staff had only to watch and control the movements of the shipping concerned."[29] While this was an adequate measure for the immediate task, it erected an additional organizational boundary within the Admiralty Staff.

The Transport Department was barely capable of keeping up with the workload imposed by global conflict, and it was completely inadequate when faced with the submarine menace:

> Various divisions dealt with aspects of the submarine problem, but none dealt with it as a single whole. Trade Division and Operations Division tended to regard it in the light of losses of men-of-war, fleet auxiliaries, and transports, and the losses in these classes of ships were admittedly few. Intelligence Division received a very considerable amount of intelligence about enemy submarines from German intercepts, but it did not go direct [to operational units], which constituted the principle force dealing with the submarine. Technical measures against the submarine were in the hands of the Submarine Committee.[30]

The measures instituted for protection of merchantmen by the Trade Division illustrate how ill-equipped the organization was. The Trade Division restricted its measurement of success to those classes of ships noted above, but was responsible for the merchant fleet as well. At the war's outset, the director of the Trade Division focused on trade routes and defensive armament for merchant ships as the primary countermeasures against submarine attacks. The director catalogued the steps taken so far: destroyer patrols in the Irish Sea and the English Channel, yacht and trawler patrols in areas with major ports (i.e., Liverpool and Belfast), and orders to merchant ships to keep a sharp lookout for submarines while maintaining full speed whenever possible.[31] By July 1915, the director noted that these measures would be insufficient if the enemy increased the number of U-boats operating against

British trade. Therefore, he suggested that trade routes should be fixed so that ASW patrols could concentrate their efforts, and if the routes were changed frequently, merchants could avoid the danger of surprise torpedo attack (in other words, unrestricted submarine warfare).[32] As a further countermeasure, the Trade Division advocated the use of "special service" ships, known as Q-Ships. These were warships disguised as merchant vessels, designed to lure a submarine to the surface by presenting a seemingly easy target and then use its concealed deck guns to sink it. First proposed early in the war, the Trade Division pressed for more of these ships well into 1916.[33]

Early in the war, the BRN had no effective method of finding and engaging a submerged submarine in use throughout the fleet. The most practicable way of destroying a submarine was to catch it on the surface. Early ASW policy reflected this. In order to protect merchant traffic, the BRN concentrated patrols at "focal points" or lanes. Merchant vessels would concentrate at these focal points, but sail dispersed between them. Smaller armed ships would do the patrolling in the safe areas, but larger ones would patrol the routes between them. Drifters—vessels best described as fishing boats but with nets designed to ensnare submarines—would lay their nets where U-boats seemed to make most of their attacks. "Armed ships" (warships or converted merchant vessels) would make themselves known to U-boats so as to keep them submerged.[34] In this scheme, merchants were left to their own devices outside of a few guarded areas, and hunting patrols would accomplish the majority of ASW tasks.

The above policy was not set in stone. The Admiralty Staff debated internally about whether or not to concentrate submarine hunters, divert trade, or even tell merchant vessels where the "safe" areas were. The director of the Trade Division argued that if merchants knew where the safe areas were, they would concentrate in those areas like sheep, resulting in chaos. Trade must be dispersed to reduce risk, and fast vessels must be used to hunt submarines. He realized that dispersal of trade necessarily dispersed the hunters, but concentrated trade made hunting too easy for the U-boats. On the other hand, the director of the Operations Division argued that ASW patrols should be assigned specific areas to patrol, and merchants should be given information as to where these safe areas were.

As a result of this debate between the two division heads, the Admiralty settled on four axiomatic principles in the defense against submarines. First, because "diversion reduces risk," trade must not follow regular routes. Second, areas around which diversion was impossible needed protection. Third, ASW patrols should be large and should work in combination with one another. Finally, patrols should be stationed in certain areas but should have flexibility to help those in other areas if needed. Vessels should also patrol between safe areas.[35] While the ASW patrols were an attempt to directly counter the U-boat threat, the merchant fleet itself was expected to absorb the punishment it received. The Admiralty concluded that any action it took regarding the merchant marine *en masse* would have a net harmful effect on trade.

The Admiralty Staff summarized the first coherent adaptations to the U-boat in a February 1917 memorandum. The BRN instituted the Auxiliary Patrol—a force of trawlers, yachts, and other small craft taken up from the civilian fleet—to hunt for submarines in home waters. It also gave as many merchantmen as possible armament in the form of deck guns and created an extensive training program in their use. Further countermeasures included offensive ASW patrols by destroyers, coastal air patrols, decoy ships, and minefields; the BRN intended for the minefields to restrict U-boat access to their hunting grounds.[36] These measures reflect the "axiomatic" principles settled upon by the Admiralty in 1915. The BRN would do its best to counter the threat, while the mercantile marine would absorb the losses.

Technical Measures

ASW is a technical business, and the BRN supplemented their countermeasures with several new technologies, not all of which were effective. The modified sweep was perhaps the most common weapon employed by the vessels of the auxiliary patrol. This consisted of floats connected by steel cables; the whole apparatus was towed behind the vessel. The floats were hydrodynamically designed to stay a measured distance apart, and were fitted with explosives. Presumably, a submarine caught within the area of the sweep would strike one of the floats and sink from the explosion.[37] Similar in nature was the indicator net; used at the entrances to ports, this was a steel net moored to the sea bottom or towed by an Auxiliary Patrol vessel. When an enemy submarine drifted into the net, the buoys' movement would signal patrol vessels of its presence.[38]

An important aspect of British adaptation was the BRN's own corps of submarine officers. Those responsible for countering the U-boat problem frequently solicited those officers for opinions on submarine performance, weaknesses, and tactics. They also evaluated the feasibility of various ASW schemes and weapons, in addition to proposing a few themselves. The BRN's adoption of zigzagging, or random course and speed changes by surface ships in order to frustrate a submarine's targeting, received aid from the submariners. Their experience in trying to sink surface ships helped make the practice more effective.[39] Given their operational experience, they were especially useful in suggesting measures that would frustrate the U-boats' efforts and lower their crews' morale. For example, submarine officers advocated placing deep minefields in areas of particularly intense submarine activity and routes, "as a particularly deadly and nerve trying form of mine warfare."[40] When the BRN eventually adopted convoy, submariners helped implement the change in policy by pointing out where the destroyer screen of a convoy would be effective or ineffective; this undoubtedly made the whole system more effective more quickly.[41]

The defensive armament of merchant ships (DAMS) illustrates further the amount of resources the BRN could expend in an ultimately misguided effort. By the end of the war, 4,000 merchant ships had been armed with 6,907 weapons; 1,712 of the weapons installed, or roughly 25 percent of the total, were lost at sea.[42] The mercantile marine was provided with 11,500 men to serve as gun crews, and 750 men fulfilled the role of gunnery instructors and installed the weapons on the ships. By comparison, most BRN destroyers had a complement of 70–90 men each.[43]

During late 1916, the BRN maintained a sizeable patrol force in the Bristol Channel area. Its mission was to protect shipping by hunting down and destroying U-boats. The force consisted of 49 destroyers, 48 torpedo boats, and 468 armed auxiliaries (trawlers, sloops, Q-ships, etc.). Despite this overwhelming presence and a force of thirteen destroyers and seven Q-ships specifically tasked with their destruction, two or three U-boats sank thirty ships in the Bristol Channel during the same time period. This situation was "typical" in the waters around the British Isles.[44]

Despite a huge outlay of resources, statistics and individual accounts point toward the futility of these devices and methods.[45] Only two U-boats sunk due to modified sweeps during the entire war.[46] Because the indicator net was not a weapon system in and of itself (it only signaled the presence of a U-boat), its effectiveness cannot be directly measured. Even early in the war, much of the Navy doubted the modified sweep's utility. The trawlers fitted with it had few opportunities to use the device, and it was difficult to ascertain whether or not they were effective, save for sending divers to investigate suspected wrecks.[47] Captain John Brooke of the HMS *Zulu* called the sweep an "ineffective contrivance" that was difficult and complicated to deploy.[48] Indeed, the Admiralty declared that "there is an unreasonable prejudice against the modified sweep" and that proof of their success (by investigating suspected wrecks, if necessary) was needed to "increase efficiency" in their use.[49] Judging by the lack of success of the sweep throughout the war, the prejudice was by no means unreasonable.

From the outset of the war, destroyers were in high demand. Besides guarding the Grand Fleet, their duties included guarding against German raids of the English coast, protecting minesweepers, escorting troop convoys, and ASW patrols. The lot of the destroyer was not an easy one. Those on station in the Straits of Dover had to contend with enemy destroyers and mines as well as submarines. British destroyers were on the defensive and operated singly, while their German counterparts left port in groups. When the two sides clashed, the British ships were not only outnumbered in the surface engagements, but also they had to be certain of their targets because all the other shipping in the English Channel was British as well.[50]

Transport escort duty required a large share of the destroyers available. Even if the government did not consider food and other raw materials essential for the British economy important enough for convoys, troopships were high value ships from the start of the war.[51] The lack of destroyers meant that even these ships sometimes went without escorts, however. The War Office complained of a transport carrying 2,000 soldiers going unescorted in its journey to the Eastern Mediterranean in late 1915. The Admiralty's response was that there were insufficient destroyers available to escort ships that distance.[52] The BRN's weakness in destroyers was exposed in the Mediterranean, where it had insufficient numbers to close down the Straits of Gibraltar, even though the local commander needed just six additional vessels to accomplish the mission in the strategic waterway.[53]

Patrols drained the BRN's resources. The Harwich Patrol had an on-paper strength of thirty vessels, but in practice had between three and twenty. The demand on these ships meant that they completed the equivalent of eight years' steaming time in just one year. The Dover Patrol had only fifteen destroyers available, nine of which were obsolescent. Of Portsmouth's ten aged destroyers, only six or seven were operational at any one time. Overall, the destroyer position was "not satisfactory." The BRN began a ship-building program to increase the number of destroyers, but the newest and best of them went to the Grand Fleet rather than to ASW forces. A surfaced submarine was no match for even the weakest destroyer; however, older destroyers had poor sea-keeping qualities and were practically useless in inclement weather.[54]

From the Admiralty's perspective, however, offensive patrolling and minefields appeared to be working. In gross tonnage, merchant ships lost peaked at 170,312 in August 1915, then declined overall until the fall of 1916, despite periodic spikes (which never rose above 152,000 tons) in ships lost.[55] The success was illusory, however. The U-boats' numbers were increasing, which would pose a serious problem when the Germans decided to intensify the war on shipping by changing their methods. While ASW patrols sank nineteen U-boats, these numbers are almost matched by the fourteen U-boats lost to accidents and unknown causes.[56] A U-boat crewman was almost as likely to be killed by bad luck, a navigational mistake, or an equipment flaw as by the BRN. ASW patrols accounted for only ten U-boats sunk in 1914–1916.[57]

Maladaptation Explained

The poor results in destruction of U-boats from the BRN adaptations in 1914–1916 beg the question: why did the BRN persist in following the same course of action after it should have been clear that it wasn't working? The answer lies in misperception. U.S. diplomatic pressure on Germany and internal German disagreements on the proper course for employing its U-boats account for the periodic decreases in merchant tonnage lost.[58] The BRN, however, perceived this phenomenon as being caused by its own actions.

As discussed above, German leaders only conceived of the war against shipping after it became clear that the decisive sea battle would be slow in coming. Hence, the German commerce war was limited in nature, bound in part by rules of war prohibiting the deliberate targeting of noncombatants. The 1914–1916 U-boat campaign still brought Germany into conflict with the United States, however, and the debate within the government regarding the direction of the effort intensified. The sinking of the White Star liner *Arabic*, which caused a loss of three U.S. lives, created a diplomatic crisis, and the German government was divided.[59] Alfred von Tirpitz, the naval secretary, led the faction that wanted the war on shipping to continue. Chancellor Theobald von Bettman-Hollweg wanted to keep and reinforce the restrictions on the campaign to assuage the United States in order to keep it out of the war.[60] The chancellor won the debate for the time being, and the U-boats became much more cautious. The Germans ran into the same problem in the Mediterranean. As soon as German U-boat operations started to take U.S. lives, another diplomatic crisis arose, and the civilian leadership was forced to restrict the campaign in those waters as well, with a consequent loss of effectiveness.[61] Finally, in April 1916, the German civilian leadership emerged from the debate temporarily victorious, and reasserted that no ship would be sunk without being searched and the passengers saved.[62] When the submarine campaign consequently lapsed in the summer of 1916, it was due to German actions resulting from U.S. diplomatic pressure, not to the countermeasures of the BRN.[63] Indeed, the United States had "whittled away" the first German submarine campaign through diplomacy.[64]

Simultaneously, the BRN had laid an extensive mine barrage off the Belgian coast. Rather than a defensive barrier protecting its shipping lanes, this minefield served an offensive purpose: to bottle up U-boats in their ports. The same day that the BRN completed the minefield, the Germans recalled their submarines as they reconsidered their strategy. The BRN took the subsequent drop in U-boat activity to mean that the minefield was effective: "Once again, coincidence of change in German submarine policy with the initiation of a new sea-barrier gave the idea that the enemy had received a serious setback."[65] The BRN had gained a respite in the submarine war, but it led the Allies to the erroneous conclusion that their countermeasures were working—a "convenient and consoling presumption." Additionally, it delayed an effective reply to the U-boat because the rationale for aggressively pursuing a more effective response disappeared.[66]

The Adaptations, Phase II: A New Organizational Structure

When Jellicoe became First Sea Lord in late 1916, he inherited the worsening U-boat situation. As the Germans renewed their submarine offensive in autumn of 1916, "the problem had outrun the capacity of Operations Division and Trade Division."[67] While high-profile targets such as warships and troop transports traveled in comparative safety, the mercantile marine had to make due with trade route instructions from the naval staff as its only protection. The U-boats had stopped surfacing as a prelude to their attack, depriving Q-ships of targets, and defensively armed merchant vessels sank when struck by torpedoes as readily as unarmed ones. Continually changing the fixed routes did not do much good either; the Germans knew there were a limited number of destinations for merchant traffic, and positioned their U-boats accordingly.[68] The artificial halts in the submarine campaign had just delayed the realization that the British response to the U-boat was fundamentally unsound.

Jellicoe's first priority when taking over the Admiralty was to address this weakness. He did this by correcting the organizational shortcomings of the staff by adding a new department, the Antisubmarine Division (ASD).[69] With the creation of this department in December 1916, the BRN streamlined its organization. ASD's duty was to supervise the movements of all vessels engaged in ASW. In addition, it was to supervise the development, material, and training for the use of all ASW "appliances." Correspondence involving the ASW suggestions, U-boat intelligence, trade routes and convoys, merchant tactics and training, and ASW devices were to be marked "ASD."[70] This department became a central clearinghouse for current and suggested submarine policy in all areas (see figure 2.2).

The second round of adaptations consisted of the convoy, together with technological advances such as the hydrophone, an early version of passive sonar, and depth charge. These advances are based on the fact that water is denser than air. The hydrophone allowed an operator to listen for noises created by a submarine and estimate its bearing from the listening device.[71] The depth charge used explosives to transmit a shockwave through the water that would puncture the submarine's hull without the charge actually having to hit the submarine.[72] Deck guns relied on the submarine being at or very near the surface for their effectiveness. When U-boats were still surfacing before their attacks, the deck guns were still nominally useful, and they accounted for a large portion of U-boat losses during 1914–1916. These guns became completely ineffective, however, when the Germans turned to unrestricted submarine warfare. With the increased availability of hydrophones and depth charge, the share of total U-boats destroyed by this combination rose from 50 percent in 1917 to 60 percent in 1918 for escort vessels, and from 19 percent to 67 percent for ASW patrols.[73]

All of these adaptations were limited in use from the outset of the war, and were not strong enough to make an appreciable difference. In order to get reinforcements safely to the Western Front, the BRN grouped troop transports into convoys from the outset of the war. Not one of these transports was lost.[74] Further adoption of the convoy was slow, however. The Admiralty accounted for the troop convoy success with the restricted waters and short routes through which they sailed. The crews of the merchant ships that served as transports also represented the cream of the merchant fleet, and therefore were able to stay in convoy formation.[75] Convoys would have been useless if the merchant vessels could not hold their stations; the convoy would become strung out and the escorts would not be able to protect the merchants. The

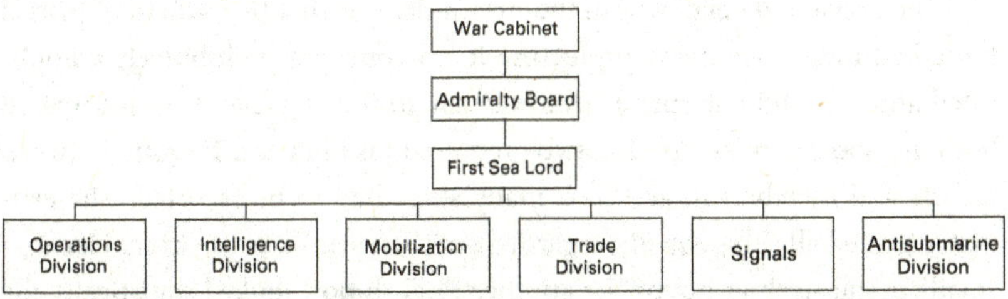

Figure 2.2
The British Royal Navy's Admiralty organization, 1916.

prevailing attitude of the Admiralty toward the mercantile marine outside of those serving in transport duty was that it was not up to the task of navigating and station-keeping within a convoy, especially over long distances and in bad weather.

Even after pressure began mounting to adopt the convoy, shoddy intelligence clouded the picture from the Admiralty's perspective. First Sea Lord Jellicoe met with a group of ten merchant captains in early 1917 in order to get their opinions on possibly implementing a convoy system. The directors of Antisubmarine, Trade, and Operations Divisions also attended. The reaction from those present to the idea of general mercantile convoys was uniformly negative. The merchant captains were of the opinion that station-keeping at night and in bad weather would be impossible due to the poor quality of both their crews and the coal used for fuel. The cure, in their opinion, would be worse than the disease.[76] Many sources use this meeting as an example of why the BRN still hesitated to adopt the convoy on a broad scale.[77] The information Jellicoe gained in this meeting, however, arguably suffered from the "garbage in, garbage out" effect. Shortly after the war, Admiralty correspondence identifies many of the merchant captains at that meeting as independent-minded coastal sailors whose conception of convoys was anathema to their way of life. Also, a significant number of captains had recently had their ships sunk out from under them while under escort. These factors led to an inordinate number of captains who had a negative opinion on convoys representing the merchant marine at a conference designed to get an honest appraisal of the practicalities of the convoy system. The results of the conference were biased, because those attending the conference were "not representative of the Mercantile Marine in any way." If they had been, the captains would have "declared unanimously for the adoption of the Convoy System. Their opinion at that stage of the War, would in all probability, have been the means of saving the Country many hundred thousand tons of shipping."[78] This is one of the factors that inhibited the quick adoption of the convoy system.

The prevailing belief within the Admiralty was that the growth of British trade had long since made protecting it via convoy a prohibitively complicated and wasteful enterprise. This was due to the fact that it overestimated both the number of escorts a convoy required (as high as a 1:1 ratio between escorts and merchants), and how many ships had to be escorted. The estimates treated all ships equally, regardless of their size or destination. Hence, a small coastal steamer bound for another English port looked statistically the same as a large ship on a transatlantic voyage. The result of these poor statistical methods was that they turned 120–140 arrivals per week needing escort into 2,400 per week. Jellicoe and the Admiralty understandably thought this an insurmountable problem, given the number of escorts available.[79]

"Short Routes" and the Eventual Adoption of General Mercantile Convoys

As discussed above, the BRN was not completely resistant to the use of convoys; the convoy (and resulting safe transit) of scores of troop transports across the English Channel and the world's oceans is a testament to this. The transition from the Admiralty's "hands off" approach to trade protection to one of convoys was a gradual one. First, it started to protect trade routes between the British Isles and various locations in Europe, which I term the "short routes." These routes included the Dutch and Norwegian trades, along with the French coal trade, which all shared certain characteristics. Close to the conflict zone, these routes were especially vulnerable to U-boat activity. Many of the vessels employed on these routes were from neutral countries. Finally, of course, the voyages were of relatively short duration, which simplified the escort problem. A cursory look at the timing of the adoption of convoys for these routes compared to that of the general convoys points to external pressures (from allies and civilians) as decisive in forcing the BRN to adopt a general convoy system. Short-route convoys were first, followed by the general routes. A deeper examination of why and how these short-route convoys were adopted reveals that, without the creation of ASD, their success would not have had the effect that it did in pushing the BRN toward a general convoy system. The mechanism by which a dedicated organizational subunit (DOS) implements effective adaptation is clearly visible here. The creation of ASD shifted the interest of the Admiralty Staff from a fixation on protection of the battle fleet to that of mercantile trade, released it from preconceptions about how to analyze the problem, and grouped the relevant expertise together.

The Dutch trade was the first to be convoyed. The route between the neutral Netherlands and the British Isles proved to be especially vulnerable to German surface and sub-surface activity. The proximity of the route to German naval bases meant that a surface raider could dash in among the merchant traffic, sink a number of vessels, and escape without a response from British patrols. Submarines also could easily claim victims along this route. This problem became apparent in June 1916. By July 26, the BRN had implemented a system of "controlled sailings," which was a euphemism for convoys.[80] The time elapsed from German attacks and pressure to start convoys to implementation of the convoy was about one month.

The Norwegian trade presents a somewhat more complicated picture. Losses suffered by Norwegian shipping in 1916 led their ship-owners' association, their insurers, and the British Board of Trade to press the Admiralty for a convoy.[81] At the end of his tenure as Commander in Chief of the Grand Fleet, Jellicoe proposed a convoy system for this route on December 3, 1916.[82] The Norwegians, however, were still suffering losses in March 1917, apparently because the BRN was not implementing the system effectively. Norwegian shipmasters claimed they saw British patrols or escorts only rarely, and many ships were still sailing singly. By April 24, the Admiralty had approved an improved convoy proposal; the first one sailed on April 29.[83] Norwegian shipping losses dropped significantly after that. Again, the time elapsed between application of pressure to convoy implementation was about three weeks. It should be noted that the demonstrative effect of the Norwegian convoys had to be limited. This is because the German unrestricted campaign was already in full swing by March, and the Admiralty had accepted the notion (if not yet fully executing implementation) of convoy in April 1917. Also, as ASD had not yet been created when the first ineffective system was implemented, the escorts were still under the control of Operations Division, with its aforementioned predilection for offensive ASW patrols at the expense of defense of trade. This may account for the paucity of escorts witnessed by the Norwegians.

As noted above, First Sea Lord Jellicoe created the ASD of the Admiralty Staff in December 1916. This new division, coequal with others such as Operations, centralized the information-gathering, analysis, research, and operations tasks concerning the submarine problem. In September 1916, shipping losses had reached 200,000 tons a month, which was unsustainable, and the current structure of the Admiralty was incapable of handling the problem.[84] The commencement of the unrestricted submarine campaign in February 1917 only made the shipping losses worse.

The French coal trade also increasingly suffered as the German submarine campaign intensified. The implications of this, however, were much more severe than the losses suffered along the other short routes. French industry depended upon British coal for its energy needs. If that supply were severed, French production would grind to a halt and the war effort would suffer. In the closing days of 1916, a meeting of British ship owners, coal exporters, and coalmine owners unanimously concluded that the coal trade should be convoyed. They saw "nothing impracticable" about convoy, and the safety of the vessels far outweighed the "little time lost" due to resulting delays in shipments. Armed merchantmen could provide the necessary escorts.[85] The

Admiralty, however, had already scheduled a conference with the French on January 2, 1917. This is important because of the timing. The Admiralty seemed to be putting off the domestic civilian pressure while simultaneously working with France to implement a convoy of the coal trade. This is evident in the endorsements of the division heads and the Admiralty's reply to the Board of Trade.[86] The director of the Trade Division was still wary of convoys. He asserted that defensive armament and adherence to trade routes should be sufficient, and that a convoy would not "necessarily protect a vessel from underwater attack." Also, convoys meant delays and congestion at ports, and participation by neutral vessels was "open to doubt" for reasons of international law. The director did not close the door to convoy entirely, however: he wrote, "it is understood that the matter is up for discussion on 2 January."[87] The director of ASD's endorsement was slightly more concrete: "when the proposals of the French have been presented and considered, action can be taken."[88] Because it was understood before the conference that France were going to propose convoy, the ASD endorsement implies that all that was required for the implementation was the French proposal, as the convoy would require joint operations between the French and the British.[89] What is curious about this endorsement is that it is dated on January 10, 1917, which was about the same time that the Admiralty was moving to approve convoys already. Furthermore, the Admiralty's response to the Board of Trade inquiry, which stated that the proposal for convoy was "under consideration," was not sent until January 16.[90] The timing of this correspondence is curious. By then, the French proposals had already been received and considered. It is hard to account for the discrepancy regarding when the convoy of the French coal trade was "under consideration," and when the decision had already been made. It seems to indicate, however, that outside pressure from British civilians was not the decisive factor in the adoption of convoy. Rather, pressure from an important ally, enabled by new BRN organization, was much more important. Indeed, the Admiralty was seemingly brushing civilian suggestions aside, even as it had made a decision for which the civilians had advocated. It is hard to account for why the Admiralty behaved in this manner; perhaps it was trying to maintain its autonomy.

The conference with the French seemed to be the decisive moment in protecting the coal trade. In attendance at the conference on January 2 were the representatives of Trade, Operations, and Antisubmarine Divisions. Commandant Vandier of the Ministry of Marine represented France. His government was primarily concerned with the flow of coal, rather than direct shipping losses. When submarine activity was detected in the English

Channel, the most reliable method of preventing further losses was to halt merchant traffic. While this stopped ships from being sunk, it had the effect of a virtual blockade on French coal imports for 30–40 percent of the time during November and December 1916.[91] The French leadership suggested that colliers should be sent across the Channel regardless of submarine presence, "provided that a system of convoys were established."[92]

The Admiralty's main objection to a system of convoy was the difficulty of including neutral ships. Many ships from neutral countries participated in the French coal trade, and those countries' neutrality would have been threatened if protected by a belligerent's warships. If the word "convoy" was officially used to describe the system, the Admiralty argued, the Germans would pressure the neutrals to withdraw from the system, drastically reducing its effectiveness in protecting the coal trade. In order to avoid the "objectionable" use of the word "convoy," Admiralty representatives suggested instructing neutral ships as to their departure time, route, speed, and destination information. A "patrol" would be in place along the designated route; the course, speed, and timing of the patrol would be remarkably similar to that of the neutral colliers. This was a *de facto* convoy, as the patrol was really an escort. While the conference attendees also recognized the difficulties inherent in "group sailings" (in maintaining formation, departure delays, etc.), they recommended the immediate adoption of these "controlled sailings."[93] The report of the conference, which is very much in favor of the French proposal for convoy, was written by Commander Henderson, the assistant director of ASD.[94]

This sequence of events begs the question: why did Admiralty resistance to the idea of convoy disappear so suddenly? The plight of Great Britain's most important ally was undoubtedly very significant. Without British coal, the war effort in France would have collapsed. The role of ASD cannot be ignored, however. First, its deputy director advocated the convoy approach outlined at the Vandier conference. Even though the French were close allies, they were still outsiders; ASD advocated for convoy from within the Admiralty. Also, while the directors of the Trade and Operations divisions were willing to experiment with the idea, they were still lukewarm to it and only willing as long as it did not increase their workloads. The presence of the newly formed ASD assisted in that regard, and the language of the endorsements of the Vandier conference report suggest that this was the case. At this point in the war, the Trade Division did not have the capacity to plan such complicated operations as those involved with a convoy system.[95] In their endorsements of the conference report, the directors of Trade and Operations recommended experimenting with convoys as long as ASD took the lead in planning them. As ASD was now responsible for coordinating patrol activity, it seemed the natural choice for arranging the "patrols" associated with the new controlled sailings.[96] ASD, however, seemed to have realized it got more than it bargained for:

> The working of the proposed scheme does not appear to me to concern the ASD in any way, but if approved to give it a trial Commander Henderson will be available to assist in working out the details.
>
> The organization will be mainly concerned with such work as classifying ships, arranging convoys, writing sailing orders, [illegible] with neutral shipping, which are obviously matters with which this Division has no concern. Even in the matter of providing escorts for the convoys this division is not directly concerned, as, under the present arrangements, escorts for the Auxiliary Patrol are detailed by DOD.[97]

It is understandable that officials will, for reasons of convenience, advocate changes in policy that will reduce the requirement for personal effort, or oppose changes that increase that requirement.[98] It appears that this was what was happening in January 1917: Trade and Operations saw an opportunity to foist these onerous tasks onto the new division, while ASD resisted them. The Chief of the War Staff overrode ASD's objections, as the division's responsibilities meshed well with implementing the new system. ASD was just nineteen days old, its responsibilities yet to be finalized, when it gained a whole new set of tasks. Indeed, the decision to adopt the general convoy system in late spring of 1917 led to the Admiralty drastically increasing the authority of ASD, especially in regards to detailing escorts.[99] As the convoy came into general use, the scope of ASD's authority grew.

While Trade and Operations viewed the controlled sailings of the French coal trade as experimental, ASD was forthright in its endorsement of the idea, if not the assignment of duties. The Admiralty's Chief of War Staff favored the general outline of the coal trade convoy system in his endorsement on January 16.[100] The first "controlled sailing" occurred on February 10, 1917. As with the Dutch and Norwegian routes, the time elapsed between pressure on the Admiralty to start convoys on the one hand, and implementation of a convoy system on the other, was just over one month. ASD was in a position to monitor the new system and evaluate its effectiveness.[101]

Unrestricted submarine warfare, which was responsible for the dramatic increase in losses on the transatlantic routes, started in February 1917. In late

April 1917, the director of ASD recommended the adoption of the convoy system to the Admiralty. Simultaneously, he also supplied information to the Secretary of the War Cabinet through channels outside of the chain of command in order to dispel the myth of the overly burdensome complexity of convoy implementation.[102] ASD thus served to clarify the submarine problem and advocate what turned out to be the most effective solution. The Admiralty then implemented convoys on a trial basis.[103] These early convoys demonstrated their potential, and the Admiralty committed to full implementation in August 1917.[104] Another source states that the Admiralty made the final decision on June 25.[105] If one assumes the later date for full implementation, the elapsed time between the pressure applied by unrestricted submarine warfare and full commitment to the convoy was about five months.

Table 2.1 illustrates the various lengths of time between the BRN receiving pressure to adopt convoy along these various short routes and the actual implementation dates of the systems for those routes. Note that the short routes each have a delay of about one month (give or take one week) between the pressure applied and implementation of the convoy. The Atlantic trade, however, which was more general in scope and larger in volume, has an elapsed time between the application of pressure (loss of shipping) and full implementation of about five months. Even taking the earlier date of April 1917 for convoy adoption "in principle," the time required for adoption in the general mercantile trade was significantly longer than with the short routes.

When the BRN finally implemented the convoy, the demand for destroyers increased. According to internal documents, the best methods of foiling

Table 2.1. The Timing of the British Royal Navy's Convoy Routes

Route	Pressure to Adopt Convoy (Cause of Pressure)	RN Implementation of Convoy	Approximate Time Elapsed
Dutch Trade	June 1916 (various incidents of surface raiding)	26 July 1916	One month
Norwegian Trade	17 March 1917 (neutral losses, shipowners' letter)	4 April 1917	Three weeks
French Coal Trade	2 January 1917 (conference with French)	10 February 1917	Five weeks
General Mercantile Trade (Atlantic)	February 1917 (unrestricted submarine warfare)	August 1917	Five months

a submarine attack were a) high speed and irregular course, b) shallow draft, and c) an "escort of small craft well disposed." If "a + b + c = safety," a decrease in a) or b) necessarily meant an increase in c) "to make up the equation."[106] Since few merchant ships had high speed or a shallow draft, the demand for escorts rose precipitously. Part of the reason it took so long to employ the convoy system was the lack of sufficient escorts.[107] Despite early, lightly escorted convoys suffering comparatively minor losses, the Admiralty still viewed concentration of shipping without sufficient escort as dangerous because it merely provided a bigger target.

The effect of the shortage in escorts should not be overestimated, however. One of the reasons the long-distance convoy routes were a more complicated proposition was that they required two separate kinds of escorts. In the open ocean, submarines were unable to chase down merchants, as even their plodding speed was too fast for the U-boats. Surface raiders, however, still posed a problem: the wide expanse of the ocean gave them room to maneuver and hide, and they had the speed to chase convoys. As raiders were usually converted ocean liners or light cruisers, they had heavier armament than destroyers. Cruisers could pick apart a destroyer escort, then have the speed to hunt down the merchant ships of the convoy. A cruiser escort strong enough to repel surface raiders was helpless against submarines, however, as these ships were not ASW-capable. Therefore, convoy escorts fell into two groups: the open-ocean escorts and the submarine danger-zone escorts. The former consisted of merchant and light cruisers, while the latter were destroyers and trawlers (the latter roughly equivalent to modern-day ASW corvettes).

Convoys would transit the open ocean under the watchful eye of the cruisers, then ASW-capable escorts would meet them as they approached the submarine danger zone around the British Isles. The BRN had plenty of cruisers and trawlers, but was under strength in destroyers. The Admiralty estimated that it required 136 destroyers and 48 trawlers to protect Atlantic convoys. It only possessed 81 destroyers, for a deficit of 55 in September 1917.[108] The convoy system in the Atlantic had been in full swing throughout the summer, however, and appeared to be working. If resource scarcity had been decisive in the decision to adopt the convoy system, it should have prevented the BRN from adopting it, or at least drastically limited its effectiveness. Also, when the deficit of 55 destroyers is compared with the Bristol Channel patrol force of 49 destroyers (not to mention the other ASW ships assigned to it), this demonstrates that the BRN had more destroyer strength available, but just frittered away a large chunk of it on the patrol mission.

By September 1917, the Admiralty had seen the positive results of the convoy system. Of 829 vessels convoyed on both inbound and outbound voyages, only ten had been lost. The advantage of protection of trade clearly outweighed the disadvantage of wasted tonnage.[109] Despite continued reliance on offensive ASW patrolling elsewhere, merchant ship losses continued to mount in those areas. As the Admiralty Staff came to this realization, they recommended extending the convoy system to other areas such as the Mediterranean.[110] The BRN's operational commands were coming around as well. Commander in Chief Plymouth wrote to the Admiralty that no ship should pass through the English Channel unescorted. Despite the best efforts of patrols and route guards, U-boats were still sinking merchant ships. He argued that the BRN should employ convoys for all traffic because the routes can be varied (making the traffic harder to find), and greater protection was possible. Submarines seemed to have avoided close contact with convoys, making their attacks less successful. Most important, Plymouth recognized that patrols were ineffective, and that patrol vessels should have been shifted to escort work.[111] This demonstrates that the organization was gradually incorporating the lessons it had learned regarding the effectiveness of convoys and the relative ineffectiveness of offensive measures. Escorts were still in short supply, however. Other commands wanted reinforcements for their patrol forces despite their ineffectiveness, and some convoys in the Mediterranean sailed without escorts.[112]

Despite the escort shortage, the convoy system was fully established by December 1917.[113] Once it was in widespread use, shipping losses slowly but steadily declined throughout the rest of the war.[114] Convoy had indeed proved to be the most effective counter to the U-boat.

Analyzing the British Royal Navy's Organizational Ability

Now that I have examined the innovation and the BRN's adaptation to it, I can evaluate its ability to organize to meet the new threat. I will begin with the effect of the BRN's ability to form a dedicated organizational subunit (DOS). The BRN suffered from the U-boat disproportionately because its organizational system was ill-equipped to deal with this innovation at the beginning of the war. Changes in the Admiralty Staff organization helped to gradually improve the BRN's response to the problem.

In terms of resources, the BRN had a plethora of technically adept personnel and materiel with which to combat the U-boat. Ocean-going escorts, however, were in short supply. It was not that they did not exist; rather, they were fulfilling other roles, namely offensive ASW patrolling and escorting the Grand Fleet. The latter was necessary, given the battle fleet's strategic importance. The former was not, as patrolling proved to be completely ineffective. In this way, resource scarcity was artificial, because it could have been redressed with reallocation. Finally, in terms of the U-boat problem, the BRN was decentralized until the creation of ASD at the end of 1916. This new DOS cleaned up the BRN's problem-solving organization, as well as providing a central clearinghouse for ASW issues. Thus, the BRN could call upon substantial resources, but suffered from poor organization for much of the war.

Creation of a Dedicated Organizational Subunit

I will examine BRN organization in two different but interconnected ways—first through the lens of the Admiralty Staff's organization, then through that of the prevailing arrangement of the fleet's operational units. This is appropriate for two broad reasons. The Admiralty Staff functioned in a very similar way to an army general staff. For much of the war, its executive powers were limited; however, it advised the First Sea Lord, who in turn made naval decisions and advised the civilian leadership (the First Lord of the Admiralty and the War Cabinet). Decision-makers within and without the BRN relied on the information it gathered and the advice it gave.[115] The internal arrangement and administrative barriers of the staff are thus directly relevant to this analysis. The BRN's operational units carried out the strategies generated by the national leadership. Therefore, we must also examine how they were arranged relative to the U-boat threat. The types of ships at sea and their operational organization are relevant here.

When examined in these two ways, it becomes clear that the German submarine innovation created a competitive adaptation requirement for the BRN, especially in the operational dimension. As stated in Chapter 1, this type of requirement leads to competition over resources, leading to waste of time and effort. When the Admiralty created the ASD in late 1917, the new subunit fulfilled the role of a DOS; this accounts for the effectiveness of the BRN's adaptation after that point.

The Prime Minister's War Cabinet provided overall direction to the war effort. The First Lord of the Admiralty (the civilian head of the BRN) and the First Sea Lord (the uniformed head) both sat on this board as well as the Admiralty Board, a civilian-military committee that formulated naval policy.[116] The Admiralty Staff provided administrative support to the First Sea Lord. No one person had control over the entirety of BRN operations, but "the

First Sea Lord was responsible for advising on preparations for war, for the fighting and sea-going efficiency of the fleet, and for the superintendence of the War Staff."[117] In practice, this meant that the Admiralty Staff had a tremendous influence on naval policy and operations indirectly through influence on the First Sea Lord. After 1917, when the First Sea Lord became in effect the Chief of War Staff, the Admiralty Staff gained direct influence through the chain of command as well. Its organization greatly influenced the British reaction to the submarine threat.

As the narrative above illustrates, from the start of the war in 1914 through the end of 1916, three subunits of the Admiralty Staff held responsibilities for combating the U-boat problem: Operations Division, Trade Division, and Transport Department. Operations Division saw the problem as destroying U-boats. Because the most efficient way of accomplishing this, destroying bases, was unavailable, they had to hunt U-boats with offensive ASW patrols. The Trade Division, which was primarily responsible for protecting the mercantile marine, limited its efforts to dispersing merchant traffic and keeping it moving. It perceived a convoy system as inefficient and too complicated to implement. The subunit charged with getting soldiers to the Continent, the Transport Department, used convoys to protect that most valuable of cargos; however, it worked in isolation from the rest of the Staff. Therefore, no one subunit advocated the general use of convoys or a coordinated response to the U-boat problem before ASD was created. The system in place through the end of 1916 protected warships, English Channel transports, fleet auxiliaries, and other "valuable ships." Neutral ships and ordinary merchant shipping, on the other hand, suffered heavily.[118] In an organization long predisposed against the general application of convoy, this arrangement turned out to be almost fatal for the merchant fleet in that it encouraged continued neglect.

The creation of ASD greatly simplified the BRN's organization. Whereas previously multiple subunits of the Admiralty Staff worked in an uncoordinated fashion and the merchant marine was all but forgotten, there was now one subunit to handle all aspects of the submarine problem. As discussed in the introductory chapter to this volume, multiple subunits trying to solve a single complex problem are at a disadvantage compared to one subunit trying to solve the same problem. The fact that the Trade and Operations Divisions, along with the Transport Department, were all responsible for countering the U-boat threat resulted in a "competitive" adaptation requirement. This is because each subunit was hard-pressed to deal with the aspect of the problem for which they perceived themselves responsible. Trade and Operations saw their missions as accomplished, while the mercantile marine, upon which

Great Britain was dependent, lay outside those mission areas and absorbed the increasingly heavy blows from the U-boats.

The discussion of the French coal trade above also highlights the fact that the creation of ASD made implementation of a new method of trade protection possible. Trade and Operations were overwhelmed by their wartime duties; the pressures imposed by increased shipping losses made them more likely to accept the idea of convoys (at least as an experiment), as long as they were not responsible for implementing it. ASD filled this gap in capacity. When it was created, the new division's work comprised that of three divisions—Operations, Trade, and Intelligence, insofar as it dealt with enemy submarines.[119] The workload of which it relieved those divisions was considerable.

By the closing months of 1916, the German submarine offensive had re-intensified. While U.S. diplomatic pressure on the Germans increased, it did not have the decisive effect it once did. The Germans had made the decision to prosecute the submarine war regardless of U.S. actions. Jellicoe, as incoming First Sea Lord, recognized immediately that the BRN could not handle the U-boat problem as it was then organized. His solution was to create ASD. The new division's responsibilities were all-encompassing with regard to the U-boat problem. With the authority to direct ASW vessels and escorts, plan technical research, and evaluate statistical data, ASD had the ability to collect vital information about the commerce war, analyze it, and retain authority to act upon it.

ASD began to put the pieces of the U-boat puzzle together almost immediately. The director, Admiral Alexander Duff, monitored the progress of the minor convoy systems starting in early 1917: the Dutch trade, the Norwegian trade, and the French coal trade. While the general mercantile trade posed a more complex and larger problem than these shorter routes, they did help to prove that a system of convoy was feasible. Also, ASD carefully evaluated the countermeasures in use: the new method of torpedo attack without warning had rendered most of them (especially the Q-ship and defensive armament of merchant ships) obsolete. Furthermore, the escort problem, while serious, was not completely insurmountable. When it came to escorting convoys, the Navy would not have to escort 5,000 vessels a week as previously assumed. This large number of vessels included all sorts of smaller ships that rarely went into harm's way, represented local traffic, or both. There were only approximately 140 transoceanic departures per week that needed to be protected.[120] ASD was the first division to put all these facts together and come to what turned out to be the proper solution. The convoy had at last gained an effective, unitary advocate in the form of ASD. Figure 2.3 illustrates how the new subunit simplified matters for the BRN.

At the operational level, the BRN anticipated Germany's use of the submarine in World War I; after all, it possessed its own submarines. What it did not anticipate was how the German Navy would use this new weapon. In a fleet auxiliary role, the submarine posed a serious but surmountable problem for the BRN's Grand Fleet. This was due to the nature of the submarine's weapons and its mode of attack. The primary weapon of the submarine was—and still is—the torpedo. This weapon was the most effective way to sink a ship for the simple reason that water conducts shock better than air.[121] Battleships were constructed to absorb a tremendous amount of punishment from enemy gunfire, i.e., from above the water line. Torpedoes bypassed this protection and delivered the shock of their warhead detonation against the thinner hull below the water line. The threat of the torpedo was not new by World War I, however. Navies worldwide constructed small, fast torpedo boats that could get under the big guns of the battleships and launch their weapons, thus threatening the dominant naval weapon system with a relatively cheap one. This was the reason that battleships mounted smaller-caliber guns in addition to their main armament, and that navies developed the "torpedo boat destroyer." This new type of vessel mounted a large number of smaller guns and its own torpedoes, and could achieve speeds that enabled it to screen the battleships and to catch the torpedo boats. Because the submarine launched its torpedo from below the sea's surface, it bypassed the established defense against this weapon, and made the battleship vulnerable once

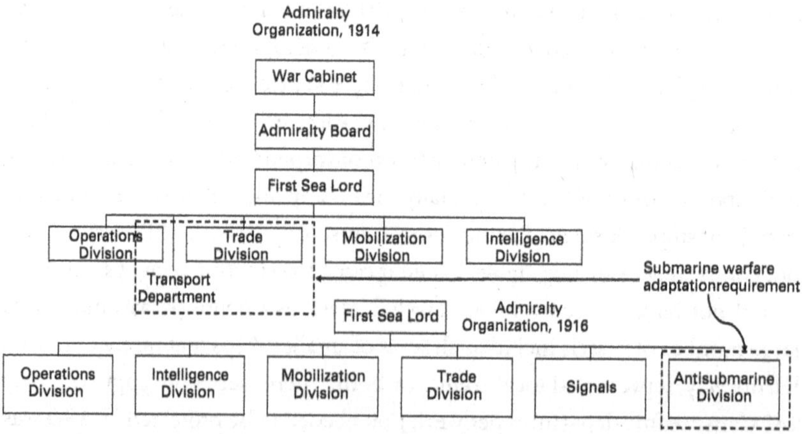

Figure 2.3
The creation of a dedicated organizational subunit within the British Royal Navy.

again. The concern within the BRN was not that swarms of submarines would sink every battleship. There were not enough of them to do that, nor were they fast enough, either submerged or surfaced, to make sustained attacks on a battle fleet at sea. Rather, the BRN was worried that a submarine "ambush" would whittle away at its battleship superiority and thus weaken it relative to the High Seas Fleet prior to the decisive battle.[122] The BRN thus had to work hard to develop counters to the submarine to protect its battle fleet.

When the Germans added the commerce raider mission, however, the submarine presented an adaptation requirement that encompassed multiple subunits of both the Admiralty Staff and the operational fleet. Not only did the BRN battle fleet have to worry about U-boats; now other subunits of the BRN had to deal with them as well. This magnified the complexity of the problem and its demand for BRN resources.

When we look at the organizational arrangement of the BRN outside of the Admiralty Staff, it consists of two subunits: the Grand Fleet, and everyone else. The Grand Fleet was the center of Great Britain's naval strategy, and the center of the Grand Fleet was its dozens of powerful battleships and battlecruisers. All other ships in the Grand Fleet, including cruisers and destroyers, fought in support of the battle line. Those cruisers and destroyers of the Grand Fleet that periodically went on ASW missions were fulfilling secondary roles in the eyes of their commanders, and had to be ready to rejoin the Grand Fleet at a moment's notice. "Everyone else" consisted of home defense and patrol vessels, the mercantile marine, and merchant cruisers (passenger liners armed for commerce raiding).

This bifurcation into two subunits is appropriate for two reasons. First, the vast majority of Great Britain's naval power was grouped around its battleships of the Grand Fleet. The philosophy of naval warfare during World War I revolved around the all big-gun ship and the decisive sea battle.[123] Accordingly, the Admiralty's first goal throughout the war was to ensure that the Grand Fleet was prepared to meet the German High Seas Fleet if it came out of its bases. In a fleet action, battleships were vulnerable to destroyers and submarines armed with torpedoes, hence the necessity of cruisers and destroyers for escort. The BRN was adamant about the need for Grand Fleet readiness in this respect, and for good reason. Without the Grand Fleet, the BRN would not have been able to deter the High Seas Fleet from coming out of its bases. Even after the convoy system was in place and destroyers were in very high demand for merchant escort work, British naval policy stated that the Grand Fleet "shall not be weakened" and a significant number of destroyers should stay with it.[124]

Second, the BRN itself viewed the U-boat problem in terms of these two subunits. The safety of the Home Islands depended on the Grand Fleet. If enemy submarines were able to whittle away at the BRN's battleship strength, the Germans could seriously jeopardize that safety. Hence, the BRN's priority for ASW measures was the Grand Fleet, and the rest of its naval assets were a distant second.

The BRN's perception leads us to examine the operational dimension of the adaptation requirement (see figure 2.4). The BRN was very concerned about the U-boat in a fleet action because it envisioned its own submarines as reconnaissance assets and agents of attrition of the enemy battle line and assumed that U-boats served similar functions, and because of some early successes of U-boats against large BRN warships. One U-boat engaged and sunk three heavy cruisers in 1914, and a battleship was sunk by a submarine-laid mine in January 1915. Submarines claimed two more capital ships during the Dardanelles Campaign.[125] The effect of this perception, reinforced by the losses in men-of-war, was that the BRN had two vital parts of its sea power to protect—the battle fleet and trade.

What did this perception mean for the BRN's ability to organize? By having to split its front-line ASW units between the Grand Fleet and the protection of trade, the BRN accentuated its resource scarcity. Convoys needed escorts in order to be effective. Even though experimental convoys demonstrated that gathering merchant vessels together under weak escort deprived the U-boats of many of their targets, the BRN still viewed such convoys as merely bigger targets. A strong escort was the only way to ensure the convoys' safety. When a convoy transited the U-boats' hunting grounds, this meant a strong contingent of destroyers. These vessels had the speed, armament, and communications to be effective escorts. Simultaneously, the U-boat's threat to

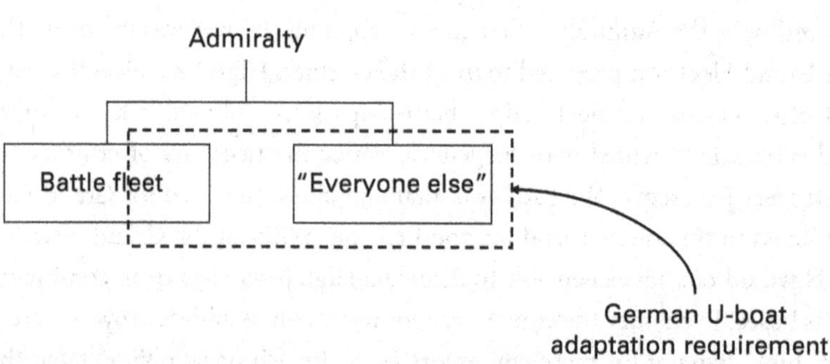

Figure 2.4
The adaptation requirement in the British Royal Navy's operational subunits.

the battle line made destroyers indispensable to the Grand Fleet. That which made the destroyers excellent for convoy protection made them vital for battleship protection. While merchant vessels plodded along at just over ten knots, the battle line sailed into combat at well over twenty. The destroyer was the only type of vessel fast enough to keep up and nimble enough to counter the enemy's torpedo-armed craft.

It is important to note the role of the German High Seas Fleet in relation to the submarine war because of its impact on the BRN's adaptation requirement. The German Navy entered World War I fully expecting a decisive action between the High Seas Fleet and the Grand Fleet. It was constructed for that purpose, and all other missions were subordinate to that role, including commerce war. Because the BRN was dominated by battleships, the German Navy had to be as well. Other types of ships in the German Navy and their missions were not completely irrelevant, but by preparing for the "worst-case contingency"—a decisive fleet action against the BRN—the Navy would be sufficient for all others. The result of this strategy was that submarines enjoyed far less support within the German Navy.[126] The number of U-boats available was not sufficient for a decisive campaign early in the war. The very gradual increase in British shipping losses reflects this. Therefore, the battleship focus of the German Navy, embodied by the High Seas Fleet, was an impediment to an effective *guerre de course*.

On the other hand, the High Seas Fleet's effect on the U-boat campaign was not entirely negative, from the German point of view. While it took resources away from the U-boat program, it also made the latter possible in two ways. First, it forced the BRN to maintain the Grand Fleet in readiness throughout the war, which was a drain on resources that could have been used directly against the U-boats. Second, the mere presence of the High Seas Fleet fulfilled its strategic role of protecting Germany's coast, and by extension, its submarine bases. I will discuss how each of these effects impacted the BRN's adaptation requirement in turn.

The presence of the High Seas Fleet as a "fleet-in-being" accentuated the resource scarcity embodied by the BRN's insufficient numbers of destroyers. In November 1917, the First Sea Lord stated that the whole of Great Britain's naval policy depended on destroyers. The Grand Fleet needed almost ninety of them in order to safely leave port. They were the most effective ASW patrol vessels because of their speed, communications, and armament. In addition, they were the most suitable craft for convoy escort duty.[127] The only reason the Grand Fleet needed to leave port was if the High Seas Fleet left its bases. Without the presence of the German battleships, the Grand Fleet would only have to sortie when it chose, and its dozens of destroyers could have been used elsewhere. Despite the fact that no German vessel larger than a destroyer ventured farther than 150 miles from the main naval base for more than a year, the Admiralty felt it had no choice but to counter the threat of the High Seas Fleet.[128] As stated in an Admiralty memorandum, "It must always be anticipated that the High Seas Fleet may come out," therefore, "the Admiralty view is that the Grand Fleet must be kept, as regards all its units, at sufficient strength to engage the High Seas Fleet."[129]

This statement was made a year and a half after the 1916 Battle of Jutland, the last major surface engagement of the war. Even though up to that point the Germans had been unable to lure the Grand Fleet out on their terms (the strength of the High Seas Fleet versus a portion of the Grand Fleet), the threat continued. Lone U-boats were ineffective against convoys, but a squadron of heavy fleet units like battleships and battlecruisers could wreak havoc on a concentration of merchantmen and their light escorts. Adding heavy escorts to the convoys solved this problem but presented a new one: the High Seas Fleet would then have the opportunity to engage the Grand Fleet piecemeal. Therefore, the Grand Fleet would have to be ready for a large fleet action that may have resulted from having to rescue a convoy.[130] Indeed, the First Lord of the Admiralty viewed the position of the High Seas Fleet *vis-à-vis* the BRN as strategically superior at the outset of 1918. The German Navy had interior lines of communication and were not saddled with the type of grinding operations, like ASW patrols and convoy escort, that the BRN was. The High Seas Fleet was still the center of the war at sea.[131] Yet, while the German fleet was idle in its ports in December 1917, U-boats had sunk over 353,000 tons of shipping.[132]

The most obvious offensive action to counter the U-boat was to destroy the bases from which they sailed, depriving them of support. Arguably, the High Seas Fleet saved the U-boats by keeping the Grand Fleet away from their bases. In response to the Admiralty request for ASW suggestions early in the war, Jellicoe endorsed "offensive measures" against the U-boats, including neutralizing their bases.[133] Absent the High Seas Fleet, the BRN would have taken such action.

An aggressive officer corps affected the BRN in another way. At the outset of the war, without the centralized coordination later provided by ASD, the BRN relied on its various subunits to counter the submarine threat separately in a piecemeal fashion. In the absence of a coherent mode of operation, the offensive mindset that was pervasive in the officer corps took over. The subunit in charge of protecting trade, the Trade Division, adopted one set of countermeasures. It instructed merchant ships to undertake tactical measures, such as sailing close to land and keeping to shallow water, and disguising themselves as neutrals to ward off submarine attack. In the event of an attack, merchants were to turn away from the U-boat at their highest possible speed, and resist the urge to aid torpedoed ships.[134] The Trade Division issued further instructions to avoid known trade routes, scatter in the event of an attack, maintain the secrecy of newly assigned routes, carry neutral cargo, and sail neutral waters when possible.[135] As noted above, the Admiralty set up the Transport Department in relative isolation from the rest of the Admiralty Staff. Faced with the same submarine menace, it instituted convoys for the safe transit of the British Army to France. Troop transports were too valuable militarily and politically to leave unescorted. Thus, two subunits faced with the same basic problem arrived at two very different solutions.

The BRN's reliance upon offensive measures produced little in the way of results. Few saw the offensive potential of the convoy, and the BRN leadership dismissed the measure as too difficult, too costly for trade, and overly defensive. Rather than deter submarines from attacking their targets, the organization wished to hunt them down and destroy them. The Commander in Chief of the Grand Fleet, Admiral Jellicoe, proposed increased numbers of Q-ships and more frequent ASW hunting patrols.[136] Q-ships roamed the seas long after they became ineffective, and ASW patrols continued until the end of the war. Offensive measures had support up and down the chain of command: "The best means of defense [against the submarine] is an *offensive* carried out by numbers of fast craft fitted with good means of communication and efficient means for destruction, who can follow and ceaselessly harry any submarine that may be reported."[137] This widely held misperception that offense was best made the BRN's resource situation much worse. It fostered an environment in which ASW patrols were vital to success, but simultaneously completely ineffective in combating the U-boat.

When the Admiralty created ASD, not only did the BRN create a DOS that fit the adaptation requirement perfectly, but it also centralized its ASW effort. This allowed the organization to evaluate its countermeasures. ASD quickly realized that the convoy was its most effective offensive ASW tool as well as effective defense for merchant vessels:

> The convoy system although of course designed as a protective measure of defense, brings in its train a useful measure of offense.... The submarine attacking the convoy is attacked in return, and, as long as a reasonable possibility of success remains, she is hunted,

> and as a convoy will always be a magnet to the enemy submarine, the possibilities of successful offensive action by the escort are always great.[138]

In matters of ASW, the BRN was too decentralized in that it had the most effective method in hand and even partially implemented, but did not take it up in a meaningful way until after losing over 7.2 million gross tons of shipping of all merchant types.[139] For much of the war, more than one subunit of the BRN tried to meet the submarine threat, with ineffective results. The BRN better positioned itself against the submarine innovation through centralization of effort. This demonstrates that creation of a DOS and centralization are closely related.

A further conclusion to be drawn from this case is that in the absence of central guidance on a particular problem, an organization is likely to fall back upon its organizational "culture" for solutions. The Admiralty provided little guidance on the U-boat problem to its operational units during the first months of the war. The officer corps' instinct for aggressive action took over, which is reflected by the actions and feedback of the BRN's operational commands. Furthermore, when the Admiralty Staff requested suggestions from the fleet, it received suggestions weighed heavily in favor of offensive action. This confirmed the Staff's own bias and partially delayed the truly effective adaptation, the convoy, because it was too "defensive" a solution.

As outlined in the previous chapter, there are alternative explanations for the sources of successful adaptation, and they can be applied to the reasons behind the BRN's adoption of convoy. These are, first, that the civilian leadership intervened to force the Navy to adopt convoy and, second, competition between multiple subunits within the Navy eventually produced the ideal solution to the problem.

The argument that the convoy idea had to come from civilians outside the BRN is the strongest alternative explanation. Civilians in the British government and industry were sources of pressure to adopt the convoy, which was especially evident in the events surrounding the French coal trade. Business and government came together in common cause in the December 1916 letter from the Executive Committee to the Admiralty. Later on, Prime Minister David Lloyd George asserted that it was his influence that finally forced the BRN to adopt the system for general trade. Faced with crippling losses in trade, he sacked Jellicoe, brought in Admiral Beatty as First Sea Lord, and removed the last barrier to convoy.[140] While this most extreme view has largely been discredited, others have asserted that some form of civilian intervention was likely more decisive than internal BRN mechanisms.[141] Jan Breemer finds that civilian intervention was more plausible.[142] Furthermore, Breemer also notes that "none of the surviving Admiralty memorandums or minute sheets in the archives hints at any consideration of convoying as a serious option."[143] My argument points out, however, that convoy was considered and applied to other aspects of the shipping protection problem. As Richard Hough writes, while Lloyd George claimed that his "peremptory action" forced the introduction of convoys, "no single figure, and certainly not Lloyd George, can reasonably claim sole credit for the belated introduction of convoys."[144] Furthermore, Lloyd George's aide, Sir Maurice Hankey, recorded in his diary that the BRN was "undertaking convoy 'on their own initiative'" three days before Lloyd George's famous "descent" upon the Admiralty to force the issue.[145] While it may be accurate to dismiss much of the consideration of convoy in Admiralty correspondence, one should also take into account that the ability of that organization to properly frame the problem and gather data before the advent of ASD was severely hampered. "Serious consideration" came rapidly after the reorganization, because before ASD, the BRN did not fully understand the technical issues of the convoy system.

This analysis also demonstrates that the united interests of the coal industry and government could not have had much of a telling effect on their own, because the Admiralty was already coordinating with France on convoy implementation by the time it responded to the Executive Committee's letter. Furthermore, contemporaries disputed Lloyd George's assertion. According to both the civilian and military leadership of the Admiralty at the time, it had already made the decision to adopt convoy by the time the Prime Minister supposedly forced it to do so.[146] As the decision to adopt convoy came from within the BRN after ASD was created, the alternative theory of civilian intervention is not persuasive.

Because Posen identifies top-down pressure as a proxy for balance-of-power factors, it is appropriate to discuss counterarguments in those terms as well. General mercantile trade was vital to the survival of Great Britain. Of all of the types of shipping to be protected, it should have been the first; however, it was not. The BRN convoyed troop ships from the empire immediately. Pressure from allies and neutrals led it to convoy the French coal trade and neutral shipping in about a month in response to each threat. As noted above, however, the delay between unrestricted submarine warfare and the adoption of convoy was approximately five months. The relative size of the threats would suggest a faster rate of adoption. If balance of power was decisive in the adoption of convoy, the biggest and most pressing threat—that to general trade—should have pressured civilian leadership, who in turn should have forced the Admiralty to adopt convoy much more quickly. What actually occurred was that the BRN took its time to execute general trade convoys. It did not have a plan ready to go, as might be expected by those looking for "serious consideration" of convoy, but neither did it respond to outside pressure.

The competition explanation for successful adaptation does not pose a substantial challenge to my theory in this case either. There was indeed competition between ideas about how to meet the submarine menace. The competition itself, however, was heavily weighted against convoy. Compared to patrolling and other "offensive" measures, convoy did not have any advocates that could realistically compete within the service. Ideally, no subunit is assured of its position, which motivates all to produce the best possible solutions. Within the BRN, however, the Grand Fleet subunit and the idea of offensive action had (and retained) their monopolies on BRN strategy. There was no meaningful competition—and thus little chance that the idea of convoy would take hold—between subunits until the creation of ASD.

I will discuss the precise effects of this resource scarcity in the next section. For now, it is important to point out that the adaptation requirement created by the U-boat provided a strong challenge to the BRN's organization in that it prompted inefficient and inadequate adaptations, and hobbled adoption of the most effective countermeasure—the convoy.

Resources

As one would expect of a great sea power of long standing, the BRN possessed a wide range of technical expertise and could call upon vast resources in support of its fleets in wartime. In response to the U-boat threat, the BRN used a staggering array of personnel and equipment. It suffered from resource weakness in two interconnected ways, however; from the start of the war, it was deficient in numbers of that most important general-purpose ship, the destroyer. Originally conceived as a guard against torpedo boats (and hence called a "torpedo boat destroyer" at the outset of the war), this class of vessel performed a myriad of tasks, not least of which was escort duty. Also, the BRN's emphasis on offensive ASW patrols as trade defense ensured that destroyers remained scarce throughout the war.

In many ways, the BRN was strong in resources. The amount of materiel, manpower, and expertise expended is in many ways staggering. These measures proved, by the organization's own admission, less than effective. It continued to expend resources on them, however, because there was nothing else available: "It is considered that the devices now in use, viz:- [sic] sweeps of various sorts, indicator net, mine nets, depth charges of various descriptions etc., have all proved their value if only as a deterrent. One of the reasons probably being that they are fitted to or worked from a large number of vessels."[147]

It is clear from the above narrative that personnel and technical expertise were also among the BRN's strengths. As the war began, it already had a group of experts dedicated to the technological problems posed by submarines in the Submarine Attack Committee. The December 1916 organizational changes took advantage of this expertise and made it more efficient by coupling it with general ASW policy. While some technical measures proved to be wasteful and ineffective—indicator nets and the modified sweep are perfect examples—this knowledge base also produced hydrophone and depth charge. Convoys made easy merchant targets increasingly rare, but these devices made it truly hazardous for U-boats to take on a convoy even if they found one.

There was one area in which the BRN was resource-poor: suitable ships for all facets of the U-boat problem—protection of the fleet, protection of trade, and offensive patrols. The ship that fit the bill was the destroyer. This scarcity was an artificial one, however, because only two of these missions were truly vital. The Grand Fleet was the center of British naval power; hence, it could justifiably call upon the lion's share of the BRN's best destroyers. Based on the Grand Fleet's war plans, a minimum of three to four flotillas of destroyers was needed to protect the battle line at sea. At full strength, each flotilla comprised twenty destroyers, meaning that somewhere between sixty and eighty destroyers accompanied the fleet when it left its bases.[148] The flotillas were rarely at full strength, but neither was the destroyer force as a whole. Even partial-strength flotillas demanded large numbers of destroyers, and the BRN's strength relative to the German Navy required that these numbers remain high. The Grand Fleet had sixty destroyers including those in refit, while the Germans could call upon a potential 156 in support of the High Seas Fleet.[149] Even though the destroyer's speed and armament made it the best vessel for the job, harassing U-boats on their way to the shipping lanes proved to be of little use. Submarines were simply too hard to find, even with hydrophones, to make patrolling an efficient response. The most efficient way of destroying U-boats was attracting them to convoys. The BRN could have dropped ASW patrol mission entirely without losing its effectiveness against the U-boat, thus freeing up a large number of destroyers for more important work. The tradition of aggressive action of the officer corps inhibited the BRN from taking this course.

Other Counterarguments: Resource Scarcity, Allied Pressure, and Insufficient Cause

It is tempting to argue that simple resource scarcity accounts for the BRN's reluctance to adopt convoy. This argument goes as follows: the organization had three broad missions: protect the fleet, protect trade, and hunt submarines. There were simply not enough escorts to make convoys an effective option, so it had to concentrate on hunting submarines. Offensive action would take care of trade protection. This argument does not work, however. First, hunting submarines proved to be completely ineffective. The technology at the time did not allow for effective hunting and destroying of submarines. Arguably, submarine-hunting should not have been a mission priority of the BRN at all. Defensive "trade protection" should have been. World War I–era ASW turned the conventional wisdom that "the best defense is a good offense" on its head. A good defense (convoy) simultaneously provided a good offense by drawing the U-boats to their hunters. The patrolling mission was unnecessary, but the BRN persisted with it throughout the war. A very large proportion of the vessels used in this way could easily have been shifted to the convoy escort role. Every escort need not have been a destroyer. Armed merchant ships, trawlers, and sloops proved capable of escorting merchants and transports throughout the war, especially in more restricted waters. Perhaps the most damaging fact for the resource scarcity counterargument is that the BRN had started to implement convoys while it was still short of escorts. Well into the development of the convoy system, the BRN estimated that it did not have enough escorts. As I have shown, the reasons for the BRN insistence on patrolling were organizational. Therefore, while the BRN did indeed suffer from resource scarcity, it was the reflection of its pervasive offensive mindset, manifested in the dominance of Grand Fleet and the bias toward ASW patrolling.

Allied pressure seemed to be decisive in the convoys of the "short routes." The French and Norwegians proved instrumental in getting the BRN to start convoys on a limited basis. The information necessary to make the final commitment to convoy, however, came from within the Admiralty. Furthermore, while the Allies seemed to provide the shock to the Admiralty necessary for it to finally try the convoy, this pressure was only decisive in the short-route question. The most that can be said of the influence of the short routes is that they had a demonstrative effect: the BRN saw the results of the convoys in place and this dispelled the skepticism about setting up a general system. This effect can be overestimated, however. The French coal and Norwegian trades traveled similarly short and restricted routes, so their success had less influence than might be imagined.[150] As I have outlined above, the time between Allied pressure applied to the Admiralty and implementation of the short-route convoys was about one month in each of the three cases (the Dutch, the Norwegian, and the French coal trades). The time elapsed from the beginning of the German unrestricted campaign—the losses from which were the primary indicator that a new countermeasure was required—to the decision to adopt a general convoy system, however, was about five months. In that time span, the BRN conducted two trial convoys to make sure that the system was feasible. The routes were longer, the escort problem was more complex, and the numbers of merchants needing escort was much larger. If Allied pressure had been decisive in the Admiralty's decision, it should not have taken so long to make a decision about the convoy system, nor should the test convoys have been necessary.

Some historians assert that the convoy was not necessary prior to 1917 because the BRN's first round of countermeasures was working.[151] Those countermeasures, however, were in fact completely ineffective. The only factors that held the German U-boat campaign back were U.S. diplomatic pressure and internal policy disagreements. The British misperceived the pauses in the U-boat campaign to be the result of their efforts. The arming of merchant ships proved to be both a short-term solution and ultimately counterproductive. As discussed above, it deprived the U-boats of the ability to surface and give merchant crews time to abandon ship because they would be shot at, so they simply remained submerged and used their torpedoes without warning. Unrestricted submarine warfare was far more damaging to British trade, especially with no convoy system in place.[152] The defensive armament of merchant ships simply paved the way to unrestricted submarine warfare. Furthermore, escorted convoys would have been even more effective against surfaced submarines, which would have been outgunned from the beginning by armed escorts. A better option for the BRN would have been to adopt the convoy for all merchant traffic as soon as practicable.

Conclusion

The BRN's organizational arrangement relative to the German U-boat innovation combined to slow its timely adaptation to submarine warfare. The early adaptations were effective only with the help of neutral diplomatic pressure and German indecisiveness. Furthermore, the adaptations merely sped up the German adoption of a much more effective form of war on commerce: unrestricted submarine warfare. The BRN was ill-prepared for this logical extension of the submarine as a weapon of *guerre de course*.

The organization of the BRN relative to the innovation tugged at its resources from two different directions: the Grand Fleet was the most important element of British naval power, but trade was Great Britain's weak spot. Both elements required ASW forces for protection. The offensive mindset of the BRN, allowed to run rampant because of decentralization, fostered the illusion that offensive patrols were right and necessary. In turn, this illusion led to the BRN wasting a huge amount of scarce resources on patrolling. In effect, it created a third mission area, ASW patrols, which was of dubious operational value.

The BRN also failed at the intelligence game in two facets: "know thine enemy" and "know thyself." It misinterpreted the halting of the first U-boat campaigns in the Atlantic and the Mediterranean as being the result of its own efforts. It also grossly overestimated the complexity of the convoy problem and its own merchant fleet's resistance to it. Both of these intelligence failures were factors in the delay in adopting the convoy system.

Thus, the German Navy's innovative use of the U-boat was more effective than it should have been. The German submarine program was hobbled by its own parent service's emphasis on the decisive engagement between battle fleets over *guerre de course*.[153] What should not be overlooked, however, is the fact that the High Seas Fleet, while depriving the submarine force of resources, preoccupied the BRN so much that the latter could not concentrate on the U-boats alone. The High Seas Fleet simultaneously held back the submarine campaign and made it possible in the first place.

A further conclusion to be drawn from this case is that in the absence of central guidance on a particular problem, an organization is likely to fall back upon its organizational "culture" for solutions. The Admiralty provided little guidance to its operational units during the first months of the war regarding the U-boat problem. The officer corps' instinct for aggressive action took over, which is reflected by the actions and feedback of the BRN's operational commands. Furthermore, when the Admiralty Staff requested suggestions from the fleet, it received suggestions that weighed heavily in favor of offensive action. This confirmed the Staff's own bias and in part delayed the truly effective adaptation, the convoy, because it was considered too "defensive."

3

The U.S. Navy and Imperial Japanese Naval Aviation in World War II

Despite Japan's military weakness relative to the United States, the Imperial Japanese Navy (IJN) deployed innovative use of air power that almost proved decisive in the Pacific War during World War II. It came in two forms. First, the IJN developed a naval air arm capable of sinking even the toughest capital ships and destroying well-protected land bases. Second, in the desperate final year of the war, it fielded the *kamikaze* to try and stop the U.S. naval juggernaut. Essentially cruise missiles with human guidance systems, the *kamikazes* inflicted massive damage on the U.S. fleet. The IJN temporarily regained some of its effectiveness. Ultimately, however, this innovation proved futile as well. How was the U.S. Navy (USN) able to prevail after suffering these two jarring surprises?

Contrary to popular belief, U.S. naval aviation was not a ready-made counterpoint for its Japanese counterpart, nor was it ready to fill the gap in naval power left by the absence of the battleships of the Pacific Fleet. So how, after a seemingly resounding defeat at Pearl Harbor, did the USN recover parity with the IJN so quickly and then exert crushing superiority soon afterward? Resource disparity only tells part of the story. Adaptability tells the other part. Resources were necessary for the USN to adapt, but it did so in a timely and effective fashion because of its organizational flexibility. This flexibility allowed the USN to turn the loss of its battle fleet to its advantage in countering the Japanese onslaught. Without the battleship in the way, carrier aviation could take the lead in countering Japanese naval power, although even the victory at Midway in 1942 still left some traditionalists within the USN unconvinced of naval air power's decisiveness. Operationally and administratively, the USN was well positioned to use what combat power it still retained after Pearl Harbor, and to support it through a long war, with both materiel and trained personnel. Finally, technological adaptations to the air threat made the USN's resource advantage in ships and aircraft all the more telling. The USN paid dearly in lives and ships for the lessons it learned at Pearl Harbor, but used those lessons to turn the tables on the IJN. Nothing more clearly demonstrated this fact than the carrier task forces that inflicted crippling losses on Japan across the western Pacific during the last two years of the war.

In the face of concentrated and sophisticated suicide attacks by the *kamikazes*—the IJN's second major naval aviation innovation—the USN had to fall back on its numbers to a greater extent. The counterinnovations developed in response to conventional Japanese naval aviation were readily transferable to the *kamikaze* problem. The problem was qualitatively different, however, and Japan was able to dramatically increase its naval air arm's striking power. Despite further technological and organizational refinements, the USN ultimately needed its numbers to prevail.

Innovation, Phase I: The Glass Hammer

Japanese naval aviation deployed an innovative use of airpower that was different from that of the USN. It was the product of the geo-strategic position of Japan relative to the United States (discussed below) in the interwar years and was designed to produce a qualitative edge over the U.S. fleet. While U.S. naval aviation shared many of its characteristics, it was qualitatively different from the Japanese version, which was initially more mature as a striking force. The United States was caught off guard by the capability and striking power of Japan's naval aviation. The U.S. version evolved into something that was more than just an enlarged version of its pre-war self or a simple emulation of Japan's. Japanese naval aviation began and spent most of the Pacific War as a support element of the battle line. U.S. naval aviation filled a supporting role during the interwar years, started the war essentially on its own, and quickly supplanted the battle line as the primary striking arm of the USN. Therefore, despite both countries possessing naval aviation prior to World War II, the Japanese version of carrier air power is indeed a case of wartime innovation and adaptation.

Geography and the political-military situation greatly affected the innovative use of air power by the IJN. As an island nation bereft of significant resources to support its industrial base, Imperial Japan needed to expand in order to compete with the other major powers. Because this expansion would bring it into conflict with the other major Pacific power, the United States, the Japanese leadership viewed war between them as an inevitability. Japan, however, would fight this war at a permanent disadvantage because of treaty limitations on the size of its navy—the 3:5 ratio to the USN dictated by the 1922 Washington Naval Treaty. More important, if it decided to break out of those limitations, the United States could dwarf its industrial output and smother

Japan in a naval arms race. These stark realities were why Imperial Japan sought to engineer a quick war in which quality would outweigh quantity.

The U.S. contingency plan for war with Japan, War Plan Orange, made this strategy viable.[1] The plan envisioned a war in which Japan (code-named "Orange") would initiate hostilities by attacking the U.S. colony in the Philippines and other U.S. holdings in the Pacific that stood in the way of resource-rich areas such as the Dutch East Indies and Indochina. To counter this and rescue the Philippines garrison, the United States would assemble its Pacific Fleet, centered on its battle line and supported by a fleet of logistical train and aircraft carriers, and sail across the western Pacific, where it would seek a decisive Jutland-style battle in or near Japanese home waters.[2]

Japan would counter this plan with the superior quality of its warships and personnel. In the final climactic battle, however, quality could have a telling effect only if the U.S. quantitative superiority was somehow reduced. Submarines would play their part by sniping at the U.S. battle line as it made its way toward Japan, whittling down its numbers. The major element in ensuring Japanese superiority at the decisive moment, however, was its aircraft carriers. Naval aviation would sink the enemy's carriers and shoot down spotter aircraft, inhibiting its gunnery and hence degrading the adversarial battle line's effectiveness. Immediately prior to the war, advances in bomb and torpedo technology made it possible for Japanese naval air units to attrite the enemy battle line itself, which would give the Japan's battle line an even greater advantage. The battleship was still the focus of this strategy, however. It started with the super-battleships of the *Yamato* class. The production of these behemoths would ensure that Japan possessed capital ships superior to those of the United States. They were the backbone of the Japanese fleet, even late into the Pacific War. The strategy was also self-sustaining, to the point of stifling further innovation in naval aviation: there must be a decisive clash of battle lines after the U.S. fleet had been weakened, therefore Japan had to build the super-battleships to have a qualitative edge. But once the IJN had the *Yamato* and her sister ship the *Musashi*, the enormous sunk cost they represented demanded that they be the center of Japanese naval doctrine well after they became obsolete.[3] They reinforced the "battleship first" mentality of the IJN at the expense of naval aviation.[4]

Both the IJN and the USN had similar visions of the future of naval warfare, characterized by two considerations. First, the surface gun was the primary means of destroying enemy capital ships. Second, aviation would exert decisive influence on a future naval battle by disrupting the opposing battle line's gunnery or damaging the battle line itself.[5] But in order to exert this influence over the decisive battle, the opposing carriers would have to be destroyed. Naval guns of the interwar period had outstripped shipboard visual targeting, and aerial spotting had become integral in both sides' gunnery doctrines. With enemy carriers present, the battleships' spotter planes were vulnerable. If one side were deprived of its spotting planes, it would be at a disadvantage. As both sides expected only *one* decisive battle, any advantage gained would be critical. The enemy carriers had to be destroyed. Therefore, even in a supporting role, Japanese and U.S. naval aviation had developed a wide range of missions, including scouting and patrol, air-to-air combat, and anti-shipping warfare using torpedo- and dive-bombers. In the late 1930s, however, the current technology did not allow for a more robust capability on the part of naval aviation. Aircraft bomb loads were not sufficient to destroy battleships. Furthermore, carriers had scant protection against even modest naval armaments. Their life expectancy when accompanying the battle line into combat was short.[6] These factors limited aviation to a strong but supporting role in the coming war. In their quest to make up for their quantitative inferiority, Japan began the shift away from the battleship in dramatic fashion. Ironically, they succeeded in forcing their enemy's navy to change this strategy, but not their own. To explain why this is the case, I will describe the Japanese innovation and catalog its shortcomings. This will provide important context in analyzing the U.S. adaptations.

In short, the future belligerents were both pursuing a Pacific Jutland. The IJN, however, was built to meet the USN on the strategic defensive, close to its bases, in a short and decisive engagement. Japan's was an unadulterated short-war strategy. While naval aviation would attrite the U.S. battle line and sink its carriers, it was not built for a long-war attrition strategy. This short-war focus affected all aspects of the Japanese innovation and sowed the seeds of its eventual demise. In contrast, the USN pursued the same type of decisive engagement, but was committed to the strategic offensive. This required that an extensive logistics train accompany the fleet on its trek across the Pacific. While its logistical train was insufficient at the outbreak of the war, the officers who would prosecute it knew of the supply difficulties inherent in Pacific operations and had already practiced solutions.[7] The USN spoke the language of the short war, but it was better prepared for a long war in many ways, which will become apparent in the discussion of U.S. adaptation below. It is first necessary, however, to analyze the Japanese innovation in more detail.

Japan's innovative use of naval aviation was defined by its improved anti-shipping techniques and technology and its concentration of naval air power. In the years leading up to the Pacific War, the IJN developed effective methods of dive-bombing and torpedo-bombing. Shortly before Pearl Harbor, it enhanced the combat power of its aviation arm by creating an unprecedented concentration of air power in the First Air Fleet.

Japanese dive-bombing doctrine was designed to deliver ordnance accurately onto a moving, heavily defended target. Until some important technical innovations in bombs and torpedoes occurred (discussed below), the primary targets were enemy aircraft carriers because battleships remained the center of the future naval battle, to be handled then by the firepower of the IJN's own battleships. In attacking their target, the dive-bombers would start off the target's bow at an altitude of ten thousand feet. After gaining speed with a shallow dive, the aircraft would nose over in a 65-degree dive, releasing their bombs at two thousand feet. Pulling up sharply, they would use speed to evade anti-aircraft fire. Committing multiple squadrons to an attack split up defensive fire and made evasion by the target more difficult. While very effective, this method was not without its costs: in order to achieve five or six direct hits, the IJN calculated that it needed eighteen attacking planes, eight of which it assumed would be lost.[8] Dive-bombing produced spectacular results. IJN bombers achieved a 90-percent hit rate in an engagement against British cruisers in the Indian Ocean. This was the result of sound tactics and intensive training.[9]

Torpedo-bombing required that the delivering aircraft maneuver to within extremely close range of the target to allow the maximum chance of success. Against an adequately protected capital ship, this was an uncomfortable proposition at best. Flying low and slow enough to drop a torpedo in the face of heavy anti-aircraft fire almost guaranteed heavy casualties among the attacking aircraft, but dropping it from farther away made a hit much more unlikely because the torpedo was unguided. IJN torpedo-bombing doctrine addressed this problem. It called for a "hammer and anvil" approach, in which multiple squadrons made their attacks from separate directions and dropped their torpedoes at less than a one-mile distance. Any evasive action the target took to avoid one attack would drive it into another.[10]

Two advances in Japanese weapons made these doctrines effective against not just lightly armored carriers but also battleships. First was the Type 99 bomb. As noted above, the major navies thought that aircraft payloads were insufficient to deliver a killing blow to a battleship. Japan solved this problem by making a 14-inch armor-piercing naval shell—normally the denizen of a battleship's main armament's magazine—suitable for deployment from a dive-bomber. A special aluminum cushion inside the warhead buffered the explosive so that it would only detonate after the bomb penetrated the target's armor (made possible by the shell's heavy casing). The striking power of the big gun was now strapped to an incredibly accurate weapons platform. It was capable of penetrating the strongest battleship's deck armor, as demonstrated by the destruction of the USS *Arizona* at Pearl Harbor, whose magazine was detonated by a Type 99 bomb.[11]

An improved torpedo was the second technological advance. While unguided, torpedoes of the day were still complicated machines that did not react well to the high-impact forces involved when they were dropped from an aircraft into the ocean. Also, in order to have a good chance of hitting the target, the pilot had to drop the torpedo at extremely close range. The initial method to counter this fragility was to fly low and slow enough to prevent damage to the torpedo. Coupled with the close release point, this had dire implications for both the pilot's and the aircraft's survivability. Advances in torpedo range and speed saved the torpedo as a viable weapon. In addition to better performance, the improved Type 91 torpedo could be dropped from a higher altitude and at higher speeds. The new torpedoes and tactics did not entirely negate the mission's hazardous nature, however, and casualties remained a big concern.[12]

Together with new weapons and tactics, the IJN undertook important organizational changes that made its air arm much more effective. Until April 1941, the Navy dispersed its carriers throughout the fleet. As carriers entered service, they would form divisions of two to three vessels, which would in turn be assigned to different fleets. This made decisive concentration of the Navy's airpower difficult, and the different carrier divisions did not regularly train together.[13] When it formed the First Air Fleet, the IJN concentrated all of its carrier divisions under one command. This organization represented the largest concentration of naval air power in the world, and helped make the IJN's collection of advanced technology and skilled pilots a most effective combat force. As Gordon Prange wrote, "Without the organization of the First Air Fleet, it is hard to see how [the Pearl Harbor] operation could have been conceived, let alone executed."[14]

Thus, Japan constructed an aviation arm that was optimized for its mission in the context of its short-war strategy. That mission, however, was not independent of the traditional Mahanian decisive clash of opposing lines of battleships.[15] Also, this arm possessed a number of critical weaknesses, which help to explain why the U.S. adaptation did not simply amount to imitation and why it was so effective. The Japanese aviation arm's shortcomings lay in organization, training, ships, and aircraft. The IJN air arm was subordinate to the surface navy. Its pilot training process was so long that the system was not able to replace wartime losses. The ships supporting naval aviation—carriers and escorts—were lacking in defensive firepower and resilience. Finally, Japanese aircraft were fragile in combat.

The organization of the First Air Fleet was revolutionary, but it still did not allow Japanese naval aviation to reach its full potential. Despite its unprecedented concentration of striking power, its organization and command were optimized for the "modified Jutland" approach mentioned above. The battle line was still the center of the fleet, and it continued to be so as late as the Battle of Midway in 1942.[16] Furthermore, aviators were not in a good position to challenge this view. Most Japanese pilots were enlisted men. A junior officer who joined the aviation ranks had little opportunity for advancement, because command of aviation units was not limited to those with aviation experience. With little incentive for junior officers to gain promotion, potential leaders were shunted away from aviation. In fact, non-aviators held most of the aviation commands of the IJN at the beginning of the Pacific War.[17] The effect of this was two-fold. First, most aviators were on an unequal footing with those who maintained the primacy of the battleship well into the war.[18] They simply had to do what they were told, as their capacity to change the status quo was negligible (as in most strictly hierarchical organizations). Second, many of those commanders using aviation assets during the war did not have a complete appreciation for the capabilities and limitations of the forces under their control—a serious weakness in any kind of warfare, but especially so in one as technical and complicated as naval combat.

As the battle line was designed for a quick war, so followed the First Air Fleet. The IJN did not include a fleet logistics train that would have enabled sustained combat operations.[19] Indeed, the driving force behind the creation of the First Air Fleet came from administrative, training, and aircraft replacement demands, "rather than from operational demands."[20]

From the eve of Pearl Harbor well into the Pacific War, there was perhaps no group of aviators more skilled and motivated than that of the IJN. The selection process for acceptance as naval aircrew was exceptionally rigorous, and their training placed the utmost demands on their physical, mental, and emotional strength. The benefit of this type of program was that it produced elite pilots and aircrew. When the time came for the decisive engagement, these men would press home the attack with discipline and skill. The cost of this, however, was that it was very hard to replace losses in combat. The training program was optimized for a short war. The IJN had no large pool of spare aviators, and "little capacity to train more."[21] Indeed, the rapid degradation of Japanese naval aviation during World War II "is attributable to the negligence of training a backlog of aviators at the outset of the war."[22] The Japanese fleet's air arm was completely unprepared for a long series of costly engagements, which the war in the Pacific turned out to be.

The First Air Fleet cannot be separated from the ships that carried it into action, and therein lay more decisive weaknesses in the Japanese innovation. To provide the fleet with the qualitative edge of the *Yamato* and other first-line vessels, the IJN shortchanged itself in escorts. This choice mainly affected the submarine campaign.[23] It undoubtedly affected the defensive screen available to aircraft carriers as well, however.[24] The force that struck Pearl Harbor went to sea with a very light destroyer defense protecting its main elements.[25] More important, the air defense provided by the escorts was seriously insufficient. Japanese doctrine allowed escort vessels to maneuver independently of the carrier while under air attack. The result was that escorts fought to preserve their own existence, not the carriers'. Also, Japanese anti-aircraft guns were small in caliber and complicated to load. They suffered from inadequate range, lack of hitting power, and a poor rate of fire.[26] Both the escorts and the carriers themselves suffered from this issue.

The internal ship organization was lacking in two ways. First, numerous officers onboard the carrier had responsibility for air defense. The complicated task of protecting the carrier from air attack thus suffered from an inefficient chain of command and distribution of intelligence. The low numbers of defensive fighters, along with poor ship-to-air radio communication, only exacerbated this problem.[27] Second, the IJN paid little attention to damage control; consequently, training, resources, and personnel dedicated to this were lacking. The IJN assigned few personnel strictly to damage control; those whose responsibility it was more likely than not had other, more pressing, responsibilities as well. Furthermore, damage-control leadership on a carrier was divided between two officers, one (the deck officer) far junior to the other (the engineering officer). The disparity in rank hampered effective cooperation because the former was rarely in a position to make demands on the latter. Perhaps most telling is that the IJN instituted no real emergency flight deck repair training.[28]

In addition, the lack of radar on Japanese ships was a serious technological shortcoming. In fleet defense, the Japanese had to rely on visual sightings by pickets and float planes. This problem caused inaccurate early warnings of incoming attacks.[29]

Perhaps the most well-known shortcoming of Japanese naval aviation was the poor survivability of its aircraft. Few have seen a Pacific War documentary without the ever-present scene of a Japanese fighter bursting into flame on U.S. gun camera footage. Aircraft of all types lacked armor protection for their crew, and did not possess self-sealing fuel tanks.[30] Despite having comparatively lighter armament than their Japanese counterparts (most often .50 caliber versus 20-millimeter machine guns), U.S. aircraft were thus able to set their enemies alight or kill their pilots with just a few hits. Coupled with the inordinately slow training pipeline discussed above, this exacerbated the IJN's increasing pilot shortage during World War II.

Lisle Rose best sums up the position of Japanese naval aviation relative to its enemies: at Midway, it was "a magnificent weapon so long as it experienced no significant aerial opposition."[31] Capable of inflicting immense damage on land bases and enemy shipping, it was equally incapable of sustaining damage and then continuing to be effective in combat. The IJN also failed to provide it with adequate logistical support. Its strength was the quick strike; given the staying power of the USN, this was also its biggest inherent weakness.

The U.S. Navy's Adaptations: From the "Day of Infamy" to the "Turkey Shoot"

It is easy in hindsight to view the rise of the carrier in the USN as inevitable.[32] The battleship, however, had remarkable staying power in the USN's thinking and doctrine. On the eve of war, battleships had a much larger proportion of ship production than carriers; this would be the case for the next several years. As mentioned above, pre-war doctrine placed carriers firmly in the auxiliary role, and the air groups that went into action in 1942 were optimized in that way. Their lack of hitting power against heavy surface units was noted throughout the first two years of the war. This was a factor at Midway (discussed further below), but also in actions in the following year. Despite numerous sorties and an elderly target, U.S. carrier aircraft were unable to sink a Japanese battleship off Guadalcanal in 1943. An aviator noted at the time, "We've got to sink it . . . or else the admirals will stop building carriers and start building battleships all over again."[33] Given that he made this statement after the landmark Battles of Coral Sea and Midway seemed to prove the ascendancy of carrier airpower, it is remarkable that the perception of its weakness had permeated down to relatively junior ranks at this late date. The lingering perception that the battleship was still the predominant instrument of naval power was all but dispelled by 1944, yet the USN's war instructions for that year still placed it at the center of all types of "major action."[34] Given that the "line of battle" ship was the center of Great Power navies for centuries, this was understandable, but how did the USN overcome this attachment and replace its battle line with carrier aviation?

In response to the IJN's innovation, the USN's air arm became not a mere imitation of the Japanese auxiliary to the battle line, but an improvement upon it in both effectiveness and independence. A combination of circumstance and design brought this about. The crisis of Pearl Harbor was USN naval aviation's opportunity to reach its full potential. Organization gave it the ability to do so.

The Run-up to Pearl Harbor

Before the war, the USN suffered from well-known resource shortages, which prohibited it from important fleet modernizations such as adequate anti-aircraft armament for its surface ships. The floodgates were opened after Pearl Harbor, but this did not become a factor until after Midway. The USN was inferior to the IJN in resources until after that battle.

Prior to 1940, the Pacific Fleet exercises were predominantly based on type—e.g., battleships would practice maneuvers with other battleships, cruisers with cruisers, and so on. The task force was "a temporary grouping of units of various type under one commander for the purpose of carrying out specific operations or missions."[35] In June 1940, Commander in Chief, Pacific Fleet Admiral James Richardson established numbered task forces, which were the "very early forerunners of the actual task forces and task fleets of World War II." These *ad hoc* organizations "facilitated inter-type training in the fleet."[36] This did not change the composition of forces or the emphasis on battleships, but enabled the fleet components to work together with what they had. Richardson's replacement, Admiral Husband Kimmel, further refined the task force system to enhance training and inter-type operability. He created three permanent task forces that would concentrate on missions ancillary to the battle line: 1) the covering force, which screened and protected the battle line; 2) the raiding and reconnaissance force, which was charged with finding the enemy and harassing its bases; and 3) the amphibious warfare forces. Task Force Two contained the carriers and their air groups. By mid-1941, the exercises these task forces carried out both at sea and on shore helped develop new doctrines and cruising dispositions. Most importantly, they gave Pacific Fleet commanders experience in commanding dissimilar types of ships together, fostering understanding of individual ship types' limitations and capabilities.[37] Hence, the USN had some foundation on which to base its improvised operations of 1942, even though it was missing its primary method of engaging the enemy, the battle line.

The task force concept relies on a system of interchangeable parts comprising the various types of forces available. The administrative chain of command of the USN was (and still is) designed as a force provider of these types. Prior to September 1942, in naval aviation,

> Combat units were divided into two categories—carrier and patrol planes. Utility planes were organized separately, and battleship and cruiser aircraft came under the type commanders for the respective class of ships. In the Pacific theater a single type was established on 1 September 1942, and given the title of Air Force, United States Pacific Fleet. This command provided a permanent administrative organization for the handling of aviation personnel, equipment, and ships. It received planes from the Bureau of Aeronautics and men from the aviation-training commands and assigned them to squadrons, provided for squadron training and the shakedown of ships, and saw that all units were maintained in a state of combat readiness.[38]

This new uniformity greatly aided in the formation of task forces, as they could now be assured that the parts were truly interchangeable.

The U.S. Navy's Administrative Strengths

U.S. naval aviation, while not considered the center of USN operations until after war began, had strong organizational support within the Navy. Recognizing that aviation required unique equipment and training support, the USN created the Bureau of Aeronautics in 1921.[39] The bureau was administratively equal to the other bureaus such as Ships, Ordnance, and Personnel. The primary responsibility of the Bureau of Aeronautics ("BuAer") was to provide the Navy with airplanes and maintain them in service. The Bureau of Personnel ("BuPers") was technically responsible for all procurement of personnel (and their orders) throughout the Navy, but in practice BuAer controlled the distribution of aviation personnel.[40]

BuAer's responsibilities were comprehensive and wide-ranging:

> The Bureau of Aeronautics makes recommendations to the Chief of Naval Operations as to the technical characteristics and limitations of

> naval aircraft, naval aviation equipment, and manufacturing sources therefore involved in the formulation of operating plans. Conducts research, makes tests, and participates with other Government agencies in the design, development, and improvement of such aircraft and equipment. Contracts for naval aircraft and aviation equipment of a technical, specialized nature. provides for plant facilities as necessary to meet production programs. Schedules, in accordance with requirements, the production of naval aircraft and naval aviation equipment and assists manufacturers in the production thereof, to the end that the various items may be delivered in the quantities, of the qualities, and at the times specified.[41]

Even though aviation played a subordinate role at the time, BuAer came to be "recognized as supreme in aviation matters."[42] As can be seen, one administrative subunit of the USN controlled most aspects of naval aviation.

The USN did not lack for air-power theorists among its ranks in the interwar years. Their presence, however, did not result in the service's loss of its air arm to an independent air force or hamper the development of the fleet's air capabilities, as it did with the U.S. Army or the British Royal Navy. The predominant attitude of U.S. naval aviators at the time was that they were "Naval officers first, aviators second."[43] Also, without battle experience to provide confirmation of air power's dominance, the aviation arm could gain little traction in the wider USN. Aviators thought of themselves as the "eyes of the fleet, which was commensurate with their striking power at the time."[44] Air advocates in the U.S. Army were well on their way to independence with their Army Air Corps, and developed their own mission of strategic bombing independent of the Army's wider mission of fighting land wars. Conversely, there was little use for strategic air power at sea. Aircraft were used to protect and scout for the fleet, not bombing strategic targets: "Naval airmen for the most part were inclined to frown on the idea of an independent air arm."[45]

In addition to BuAer, naval aviation benefited from support in the Chief of Naval Operations and the Secretary of the Navy's offices. The Deputy Chief of Naval Operations (Air) was responsible for "the preparation, readiness, and logistic support of the naval aeronautic operating forces included within the several fleet, seagoing forces, and sea-frontier forces of the United States Navy, and with the coordination and direction of the effort to this end of the bureaus and offices of the Navy Department."[46]

Immediately prior to World War II, naval aviation gained a substantial boost in the Secretary's office: "The office of Assistant Secretary of the Navy for Air was revived. Artemus L. Gates was appointed to the post, and served throughout the war. His experience as an aviator in World War I and his business and financial knowledge proved invaluable during the war."[47] The assistant secretary's responsibilities were the supervision and administration of naval aviation, as well as the coordination of air activities with other departments and agencies.[48]

In contrast to the IJN, the USN made aviation a viable career option for its officers. Naval aviation was not a dead-end career. There were plenty of opportunities for promotion and advancement for naval aviators. Furthermore, the complexity involved in running aviation commands, such as carriers and naval air stations, required aviation-qualified officers. Aviation operations were complicated and important enough to do so, even if they played only a supporting role in the fleet. Naval aviation had a strong civilian lobby during the interwar years, and in 1926, Congress, through the Morrow Board, mandated that aviators command aviation units. Since the specialty was so new and there were no naval aviator officers of sufficient rank to take over these commands, the Navy's solution was to send senior officers through flight school.[49] For example, the future Chief of Naval Operations, Ernest J. King, became a naval aviator after his fiftieth birthday. The influential head of the Bureau of Aeronautics, Admiral William Moffett, promised the ambitious King command of a carrier if he became an aviator. King accepted and subsequently took command of the prestigious *Lexington*, a converted battlecruiser and one of the largest carriers in existence.[50] King commanded *Lexington* during the years that naval aviation began to take shape, and used his ship aggressively in fleet exercises.[51] King was not unique; many other senior officers in the USN went through flight school and emerged either as aviation converts or with at least the expertise necessary to effectively employ carriers in war. Pilots developed and disseminated tactical doctrine for the USN air forces on an *ad hoc* basis. Competition between the Atlantic and Pacific Fleet aviators, as well as between Navy and Marine Corps pilots, produced many tactical advances: "A lot of it was disseminated just by people getting together and sitting around a cup of coffee. . . . It was just everybody trying to outdo everyone else."[52] Thus, while the IJN held the advantage in the quality of its pilots, the U.S. officers who would eventually deploy carrier air power in the coming war were generally better informed and trained with regard to its use than their Japanese counterparts.

Naval aviation thus had substantial representation and support throughout the USN's bureaucracy. When the demands of war required that capacity be dramatically expanded, its organizational framework was already in place. The IJN demonstrated the superiority of naval aviation in dramatic fashion, but was not fully ready to assume the role that this demanded. U.S. naval aviation, on the other hand, was ready for expansion and centrally led.

The Pacific War

Admiral King became operational head of the USN as Commander in Chief, United States Fleet (COMINCH), on December 18, 1941.[53] As the relationship between the Chief of Naval Operations (CNO) and the COMINCH was ill-defined and problematic, King was able to convince the president to combine both positions in March 1942. With the new position, King gained unprecedented power. The new order

> made King the most powerful naval officer in the history of the United States. As COMINCH, King was directly responsible to the President and was the principal naval advisor to the President on the conduct of the War. King's duties as CNO would be the preparation, readiness, and logistical support of the operating forces, as well as the coordination *and direction* of the bureaus and offices of the Navy Department.[54]

The implication of this was that not only could the head of the USN exert a great deal of influence on naval policy, but also that the person in whom this power was vested was a naval aviator. The various bureaus of the USN were notoriously independent and hard to control by any one individual. The new order did not make King complete master of the bureaus, but gave him the ability to replace their heads if they did not act according to his design.[55] Because King was a qualified naval aviator with experience commanding a carrier at sea, these unprecedented powers would bode well for naval aviation throughout World War II.

In 1944, Admiral King outlined the phases of the Pacific War, which was then in progress. His phases roughly correspond with U.S. naval aviation's stages in its evolution from auxiliary to the focus of U.S. naval power projection. The defensive phase of the Pacific War spanned the period from Pearl Harbor to the Battle of the Coral Sea. In this phase, the USN was engaged in protecting its home shores and lines of communication.[56] In its relatively embryonic state, carrier aviation during this period executed "pinprick" raids against Japanese outposts. The most important event of this period, however, was the Coral Sea action. Famous as the first "pure" duel between aircraft carriers, it was a haphazard affair, reflecting as many shortcomings in U.S. carrier aviation as it did advances in the new type of warfare. In King's following "defensive-offensive" phase, U.S. operations were overall defensive in nature but included some offensives.[57] The signal event was the Midway action. Midway was similarly haphazard, but with a greater concentration of naval air power than at Coral Sea on both sides.

The "offensive-defensive" was the phase immediately following the U.S. seizure of the initiative following Midway, but a time in which it still needed to protect its gains from Japanese counterattack.[58] The USN further optimized the fleet for carrier operations and learned important lessons at actions in the Eastern Solomons and Santa Cruz. Despite heavy losses in both carriers and aircraft through the time of the Eastern Solomons and Santa Cruz battles, replacements were on the way. At a number of points during 1942–1943, the Pacific Fleet was reduced to one active carrier.[59] The Japanese were unable to exploit this weakness, and the USN managed to recover from these shortages on numerous occasions by repairing and replacing its carriers. Finally, in the "offensive" phase of the war, the USN was able to attack at places of its own choosing. This phase coincided with carrier aviation's maturity as a striking force.

U.S. Naval Aviation Comes of Age

In the months following the outbreak of war, the center of U.S. naval power—the battleships of the Pacific Fleet—lay at the bottom of Pearl Harbor. In order to keep Japan off balance and bolster morale, the USN used what it had left—the carriers—to raid its island bases. Carrier task forces roamed the Pacific, using their air groups to inflict some damage on the enemy. These operations exposed weaknesses in U.S. naval aviation. Namely, the mix of aircraft onboard the carrier was inadequate. The number of fighters, needed for protection of the carrier and for the attack aircraft, had to be increased. Also, the air wings possessed an insufficient amount of reserve aircrew. Finally, the USN possessed unreliable torpedoes.[60] In fact, the United States entered World War II with the "worst aerial torpedo of any navy," which hindered it on a number of occasions when it caught Japanese ships at anchor.[61]

The first two major carrier battles of the war were Coral Sea and Midway. They straddle Admiral King's first two phases of the war, but occurred within weeks of each other. Together, they imparted the same lessons for the USN, so I will discuss them together.

The Battle of the Coral Sea marked the first time opposing fleets fought each other without their surface vessels coming into visual contact.

Underestimating U.S. carrier strength, Japan sent an invasion force southward toward southern New Guinea, threatening Australia. It supported the invasion with only two carriers, and the United States sent its own two-carrier task force in response. While it suffered a tactical defeat in the loss of one carrier and the damaging of another, the USN was able to stop the Japanese advance in a strategic victory. At Coral Sea, both sides mishandled their air attacks, and both air defense and fighter direction were inadequate.[62] While the loss of the *Lexington* demonstrated the vulnerability of carriers to fire in general, the resiliency of the *Yorktown* during this battle and subsequently at Midway demonstrated that USN damage control and repair practices were already quite strong. The damaged Japanese carrier was out of action for months, while *Yorktown* participated at Midway despite heavy damage sustained at Coral Sea.

The details of the Battle of Midway are well known, so a quick summary of this decisive action will be sufficient. In order to strengthen its defensive cordon around the Home Islands, which had been dramatically undermined by the recent Doolittle Raid on Tokyo, the IJN devised a plan to bring the USN into a decisive battle. By drawing the remains of the U.S. Pacific Fleet into the defense of the island outpost of Midway, the sentinel of the Hawaiian Islands, Japan hoped to finally destroy it. Having broken the Japanese codes, the U.S. Pacific Fleet decided to engage in the battle, but on its terms rather than Japan's. This time, the concentration of air power was much greater on both sides. Facing the four Japanese carriers were three U.S. carriers in two task forces. After a furious game of cat-and-mouse, Japan struck the island's air base, inflicting heavy damage, but was also sighted by U.S. search planes. Raymond A. Spruance and Frank J. Fletcher, the task force commanders, immediately launched a wave of torpedo- and dive-bombers escorted by fighters. Unfortunately, the fighters and bombers were unable to affect a rendezvous and the strike was uncoordinated. The torpedo bombers attacked the Japanese fleet alone, with disastrous results. Their sacrifice, however, pulled the Japanese CAP (combat air patrol) out of position, and the dive-bombers found the Japanese carriers virtually undefended. Numerous hits on the flattops ignited onboard fuel and ammunition, turning three of them into burning wrecks. The fourth carrier launched its own strike against the U.S. fleet, damaging one carrier, the *Yorktown*, so badly that it was later abandoned.[63]

Stripped of his air power, the Japanese admiral hoped to lure the USN into a night action with his battleships. With his air groups spent and no heavy surface units of his own, the U.S. admiral did not fall for the trap and retired eastward.[64] In a revealing incident, U.S. aircraft were able to pounce on two Japanese heavy cruisers that lingered behind their main fleet because they had collided with each other. After several strikes with all available aircraft, the cruisers were damaged, but not sunk. U.S. bombs were not effective enough against heavy armor, and there were not enough torpedo planes to finish them off. One got away when both should have been sunk in short order.[65] A revealing BuAer memo stated that "had the 1000 lb. armor piercing bomb under development been available for dive bombers, fewer of the many ships that were hit would have escaped; and fewer hits would have been needed to destroy the carriers."[66]

The Coral Sea and Midway actions, fought within a short time span, together imparted some valuable lessons, upon which the USN acted. In order to become a more potent offensive threat, carrier air groups needed reformulating. It was clear that there were not enough fighter aircraft embarked, and they suffered from misuse. At Coral Sea, the fighters ended up divided and intercepted incoming Japanese aircraft piecemeal. The pilots often received little information about the threat they faced, not because it was unavailable, but because those onboard ship who had important information, such as the probable nature and size of enemy contacts to be intercepted, neglected to share it at critical stages of the battle. As for the strike elements of U.S. air groups, they needed tactical improvements as well. At Coral Sea, U.S. dive bombers could have achieved better results by better coordinating their attacks. During that battle, pilots concentrated their attacks on some ships while leaving others untouched. Strike packages needed an officer in tactical command who could direct and coordinate the attack. Furthermore, the dive bombers made serious tactical errors that would have proved disastrous if Japanese anti-aircraft fire or CAP had been more competent. Specifically, the *Yorktown* dive bombers orbited the Japanese fleet before attacking, rather than using the *Enterprise* air group's doctrine, which closely emulated the more sophisticated Japanese one described above. Had Japanese aircraft or anti-aircraft gunners noticed the vulnerable orbiting formation of U.S. dive bombers, the U.S. planes would have suffered grievous losses. Effective doctrine needed to be more uniformly promulgated throughout the fleet.[67] The air groups at Midway inadvertently sacrificed their primary anti-shipping element, the torpedo bombers, in order for the dive bombers to get through to the enemy carriers. They needed more and better-coordinated fighter escort.[68]

After Midway, the Pacific Fleet took steps to rotate pilots out of combat theaters to boost morale (pilots could calculate their dwindling odds of survival as well as anyone), and to enable the veterans to impart their knowledge to new pilots.[69] This ensured that the pilots did not burn out from too much combat, and that their valuable combat experience would not be lost if they were shot down. Training needed to be increased, and replacement carrier air groups needed to be ready to provide reinforcements. Despite the great disparity in the length of training time between U.S. and Japanese pilots, the former were still able to achieve a favorable kill ratio against the latter.[70]

The types of aircraft the USN possessed also needed improvement. The F4F Wildcat fighter was inferior to the Japanese A6M Zero in almost all facets, except for the protection it gave its pilot and fuel tanks. These advantages proved to be important, however. From the *Yorktown* air group at Coral Sea, nine planes returned with one or more punctured fuel tanks, but only one caught fire or leaked. The armored seat of at least one plane stopped bullets, possibly as large as 20 millimeter.[71] In addition, the Devastator dive bomber was obsolescent. Replacements for these models were already in the pipeline. Their accelerated wartime development and production meant that models such as the TBM Avenger torpedo bomber and F6F Hellcat fighter would soon arrive.[72]

The problems encountered were not limited to equipment, however. The uncoordinated nature of the Midway strikes was the result of the pressing need to strike first with severely limited information. Better search techniques and training were needed. Land-based patrol squadrons needed to be upgraded so that valuable attack aircraft from the carriers were not dissipating their strength by taking up the search mission. Furthermore, when under attack, the shipboard fighter direction needed much improvement. A number of lapses due to equipment failures, inexperience, and poor organization allowed some Japanese aircraft to get through when they should not have.[73]

The USN also incorporated successful methods into its doctrine as a result of these two important battles. Despite the eventual loss of the ship, the *Yorktown*'s air-defense doctrine proved effective and was promulgated throughout the fleet. The doctrine called for a single air defense station under one officer, who maintained control over the lookouts and anti-aircraft armament. Lookouts and gunners were responsible for their assigned sectors, but if no enemy aircraft were present in a particular sector, they could engage aircraft outside of it. This allowed the ship to counter the Japanese method of attacking from multiple directions at once.[74] Beyond the carrier itself, the disposition of a circular screen protecting it was found to be most effective. All ships were to maneuver with the carrier as a guide and protect it: "all that the screening ships do, as regards to anti-aircraft, is to increase the gunfire of the carrier."[75]

Pilot losses at Coral Sea and Midway were heavy for the USN, but did not cripple their carrier air groups. Before Pearl Harbor, the USN suffered from a critical pilot shortage. Both the paucity of aviators and the inexperience of those already in the service "seriously jeopardized" naval aviation's effectiveness.[76] With war approaching, the USN had ramped up its pilot training program. It planned on supporting a 15,000-plane force by 1944. The demands of training new pilots, however, as well as commissioning new air stations and squadrons, meant that naval aviators would be in short supply in the near term. Despite, and in part because of, the new training program, the USN predicted that "this situation will get worse before it gets better."[77] At the beginning of the Pacific War, the USN was still dealing with the shortage.[78] Immediately following Pearl Harbor, the USN further expanded its training program from a capacity of 800 to 2,500 trained pilots per month. By mid-1943, the USN was training 20,000 pilots per year.[79] Also, it was able to husband its manpower through better protection of its pilots in their aircraft (through self-sealing gas tanks and armored cockpits) and rescue of downed airmen.[80] As with most interactions between nations, relative capability is more important that absolute capability, and while the USN was short of pilots during the opening months of the war, it outperformed the IJN in protecting and replacing them from the beginning.[81]

Shipbuilding: The U.S. Navy Shifts Priorities

Shipbuilding decisions reflected the result of Coral Sea and Midway and reinforced the carrier's newfound dominance. As the strategic situation in the Pacific changed month by month, so did the USN's shipbuilding priorities. The Bureau of Ships was the organization within the USN that was responsible for the design, development, and awarding of building contracts; decisions about what types of ships to produce, however, came from the president and the CNO/COMINCH.

The USN established the Bureau of Ships (BuShips) in 1940, streamlining the previous archaic organization, which split ship construction responsibilities between the Bureau of Construction and Repair and the Bureau of Engineering.[82] In the two years prior to this reorganization, the USN's rearmament program drastically increased the number of ships under construction. The new organization was needed to meet the demand. Battleships increased from two to seventeen, carriers from three to twelve, and cruisers from ten to fifty-four.[83]

The new bureau received financial appropriations from Congress, but the CNO was responsible for planning ship construction in cooperation with the General Board. Internally, BuShips would develop a preliminary design and specifications upon which basis shipyards would bid for the contract to build the ship. Decisions as to the priority of ship type were made outside of the organization. Master precedence lists guided the types of ships for which BuShips awarded contracts.[84]

The demands of global war and the changing strategic situation meant that ship production decisions became centralized. BuShips merely carried out the orders from above:

> The responsibility of proper ship design to accomplish certain objectives being vested in the Chief of Naval Operations, only that office knew the complete story. . . . If carriers proved imperative for future operations, another battleship or two would serve little or no use. . . . The Bureau of Ships concerned itself solely with executing the directives of the Chief of Naval Operations and the Secretary of the Navy.[85]

When Admiral King became CNO/COMINCH in December 1941, his influence on what ships to produce was much greater than that of the Secretary of the Navy. The latter was more often than not a messenger between the president and Admiral King. President Franklin Delano Roosevelt, a former Secretary of the Navy, did override some of King's decisions, notably on building landing craft and escort carriers (to which King objected), but King was the president's foremost advisor on naval matters.[86]

BuShips received these directives in the form of precedence lists, which established shipbuilding priorities. The precedence lists had seventeen groups, designated A, B, and C, and 1–14.[87] Not only does the BuShips official history say that the CNO/COMINCH was instrumental, but also King's influence can be deduced from these lists, as they changed nine times from the outset of the war until October 1943; the changes coincided with both the overall U.S. strategic plans and changing fortunes at sea; and King remained in close consultation with Chester Nimitz, the overall commander in the Central Pacific, and therefore was apprised of the nature of operations there and what the operational forces needed. Immediately after Pearl Harbor, battleships were at the top of the precedence list in Group 1. Carriers were in Group 2. This order stayed the same in the March 1942 list. In June 1942, the list changed again, giving priority to destroyers and destroyer escorts for the anti-submarine campaign in the Atlantic. Landing craft also were perennially high on the lists, coinciding with preparation for the landings in North Africa. In September 1942, however, after the landmark victories featuring naval aviation, carriers occupied Group B, the second highest. Carriers oscillated between Groups B and 1 through 5 until 1943. Conversely, battleships declined in both numbers and place on the list, with only one ship staying in Group 2 through October 1942. Thereafter, battleships fell as far as Group 8, including the most advanced dreadnoughts of the *Iowa* class, which were never higher in precedence than carriers after the June 1942 list.[88]

Discussions between Nimitz and King were relatively frequent, and meetings between the two shaped how the USN would prosecute the war: "Carrier warfare, amphibious warfare, and mobile logistical support would predominate."[89] The changes to the shipbuilding precedent lists did not occur in a vacuum; King made them as he and Nimitz worked to adapt the USN to the changing nature of naval combat.

The U.S. Navy Catches Up

The Battles of the Eastern Solomons and Santa Cruz came about as a result of Japan's attempt to reinforce its position on Guadalcanal. At the Eastern Solomons, Admiral Isoroku Yamamoto saw another opportunity to lure the United States into a decisive battle. He sent two battleships, three carriers, and their escorts to cover a troop convoy of reinforcements for Guadalcanal. The USN opposed this force with two carriers and a battleship plus escorts. It came out of this action the victor, sinking one carrier for no losses. One of the U.S. carriers was severely damaged, and only good fortune prevented another successful Japanese strike; however, the Japanese planes that inflicted the damage suffered mightily at the hands of more effective and better-coordinated antiaircraft fire and a greatly increased CAP.[90]

The Eastern Solomons action revealed that while U.S. carrier aviation had improved, it still suffered from critical weaknesses. The most glaring was in search operations. Both adversaries supported their fleets with shore-based patrol planes, but Japan was much more effective in reconnaissance and shadowing operations. Japanese search planes could locate the U.S. fleet, keep tabs on it, and direct attacks to it much more reliably than their U.S. counterparts.[91] During the action, the USN sent a strike toward where it thought the Japanese fleet would be. The target, however, had changed course in the early afternoon. A search plane noted the course change and sent a report, but the report did not reach fleet commanders until nightfall, long after the strike was abandoned because it could not find its target. The shore-based aircraft had a different chain of command than the task force at sea, and were often Army Air Corps planes.[92] The USN needed shore-based search planes controlled by the fleet. Using carrier air group planes for search duties, while more efficient in providing task force commanders with information, sapped the combat power of the air group because those aircraft were unavailable for combat.[93]

At Santa Cruz, Admiral Yamamoto again took another opportunity to force the United States from Guadalcanal. With his four carriers, he was able to sink one U.S. carrier in exchange for two damaged. It was a tactical victory for the IJN, but they had lost more of their irreplaceable aircrew and did not end up forcing the United States off Guadalcanal. Improved CAP and air defense over the U.S. task force exacted a heavy toll on Japanese aircraft.[94] The increased ratio of fighters in air groups "paid ample dividends in Japanese planes shot down."[95]

The desperate naval battles surrounding Guadalcanal helped the United States forge its naval aviation into an increasingly effective offensive force. The first round of adaptations, including increased fighter complements in the air groups, improved air-defense doctrine, better and more numerous anti-aircraft guns, and better attack aircraft started to make an impact. It was not the battleships' big guns that proved invaluable in these actions, but their small ones. The dozens of 20- and 40-millimeter cannons that these ships carried claimed scores of Japanese aircraft, and the battleships themselves drew many of them away from attacking the carriers they escorted.[96] While battleships were generally about three knots slower than the carriers, the former proved so valuable that ship captains and admirals alike deemed the slower speed worth the risk in order to keep the battleship in formation.[97]

The Technological Edge

As it adapted, the USN developed a technological lead over the IJN in a number of important ways. In the carrier battles of the Pacific, this manifested itself in radar, Identification Friend or Foe (IFF), and anti-aircraft guns. Both sides had radar, but the United States was quicker to equip its ships (and later its night fighters) with it.[98] Rather than depending exclusively on visual sighting by patrol aircraft and picket ships, U.S. ships used radar for surface search and, more important, to detect incoming enemy aircraft. It made fighter direction possible, and fighter aircraft did not have to loiter above their task forces unnecessarily and waste fuel. They had a much better idea of when the enemy would arrive. Radar was still a new technology, however. It was evident that radar operators needed more training, as ships first acquired many attacking Japanese planes visually, despite the presence of radar.[99]

Fighter direction had improved from the Coral Sea and Midway period, but was still less than satisfactory.[100] It was only as strong as the personnel assigned to the task. The directors often vectored the fighters into combat at a disadvantage, e.g., at a lower altitude than their adversaries. The pilots also shared some of the blame. The radio discipline during the Eastern Solomons action was poor. As superfluous chatter flooded radio frequencies, it drowned out important radio calls and hampered interceptions.[101] At Santa Cruz, the *Enterprise* fighter direction officer was new to the job. Tasked with coordinating defense of both his ship and the *Hornet*, he concentrated on defending his own ship and sent insufficient numbers of friendly aircraft against those attacking the *Hornet*. A senior officer on the *Hornet* attributed the ship's eventual loss to this.[102]

The loss of the *Hornet*, while a reverse for the USN, also demonstrated the survivability of its carriers. Japanese aircraft were able to hit the ship with five bombs and three torpedoes. Additionally, two aircraft crashed into the ship intentionally, apparently after they sustained enough damage to be shot down anyway. The damage these hits inflicted and the fires they started forced the crew to abandon ship. Towing the hulk back to friendly waters proved to be impossible for U.S. forces. Because the ship was then vulnerable to capture, the fleet commander decided to sink it intentionally. Approximately thirteen torpedoes and 350 rounds of 5-inch and 40-millimeter ammunition were expended by the escorting destroyers before the ship went down.[103] *Enterprise* demonstrated its toughness as well. Despite three bomb hits at the Eastern Solomons, it was steaming at twenty-four knots and recovering aircraft mere hours later.[104] It was resiliency such as this that gave the USN the clear advantage in aircraft carrier quality.

Closely related to radar and improving its efficiency was IFF. In use since the start of the war and steadily improving, the IFF system sent an interrogating signal to a receiver on friendly aircraft, which would in turn send back a coded response. Receipt of the correct code told the radar operator that the blip on his screen was friendly. In the uncertain and confusing atmosphere of combat, the ability to ascertain immediately if an incoming radar contact was friendly or not cleared up some of the "fog of war" and further husbanded fighter resources.[105]

After the Pacific War began, the USN took on the otherwise neglected problem of anti-aircraft weaponry on its ships. The pre-war 1.1-inch machine gun was a flop and needed to be replaced. It had neither the rate of fire nor the hitting power required to shoot down attacking aircraft effectively.[106] Three weapons were the backbone of this effort: the Oerlekin 20-millimeter machine gun, the Bofors 40-millimeter auto cannon, and the 5-inch dual purpose gun. The USN installed dozens of the 20- and 40-millimeter weapons on all of its large ships, and all combat ships had a battery of 5-inch guns. They formed a layered defense around ships and task forces. The CAP was the first layer of defense, but the 5-inch guns could lay down radar-guided barrages.[107] When attacking aircraft reached visual range, the 40-millimeter guns, in double- and quadruple-mounts, would open fire. Finally, the 20-millimeter guns, mounted singly wherever there was room to fit them, would engage any aircraft that got through. The rate of fire and stopping power of these weapons far outstripped anything the Japanese had, producing an extraordinary level of fire that made attacking U.S. ships very hazardous.

Later in the war, the 5-inch guns became even more effective with the introduction of the proximity fuse.[108] Prior to this invention, a shell had to hit its target in order to detonate, or it was detonated through a preset timed fuse. A proximity fuse contains a small radar transmitter-receiver in the nose, which detonates the shell when it detects a contact. Put simply, the proximity fuse turned a near miss into a hit. The fuse entered service in January 1943. Attacking U.S. ships became even more hazardous for the already-taxed IJN.

The Pinnacle of U.S. Naval Aviation

Along with improvements in ordnance of all types, carrier doctrine also evolved. At the early Battles of Coral Sea, Midway, and the Eastern Solomons, carriers tended to operate widely separated from each other. Many captains feared that because carriers were vulnerable to catastrophic damage caused by relatively few hits, concentrating them posed an unacceptable risk of losing them all to one strike. By late 1942, this attitude was changing among task force commanders. Concentration enhanced both the offensive and defensive power of the air groups and the carrier's escorts. After the Eastern Solomons, numerous battles had proved to the USN that carriers operated more safely and effectively together rather than separated.[109]

Coupled with improved doctrine, technological adaptations to the air threat helped extend the USN's growing lead over the IJN in carrier warfare. The culmination of these changes was the fast carrier task force. In late 1943, these task forces started roaming the Pacific almost at will, inflicting great damage on Japanese bases and shipping. The battle that showcased how far

U.S. adaptations had advanced carrier aviation, however, was the Marianas action in 1944.

Determined to stop the U.S. advance toward Japan, the IJN sent its reconstituted carrier force to confront Task Force 58 in the Marianas Islands area. The IJN had reorganized itself, finally placing the carrier at the center of its doctrine rather than the battleship.[110] While numerically inferior, the IJN carriers possessed full-strength air groups, and 1,500 ground-based planes supported them. The Japanese admiral, Jisaburo Ozawa, planned to use his planes' superior range and island bases to get a jump on the United States. While the IJN was able to launch four large raids against the U.S. task force, U.S. pilots had nearly twice as much training and better planes than their opponents. They shot down almost 350 Japanese aircraft for the loss of only 30. Task Force 58 was able to accomplish this while devastating many of the Japanese airfields that supported the operation. Despite the lack of U.S. counter-strikes on his carriers, Ozawa's force, and the IJN as a whole, had lost its naval aviation for good.[111]

The Pacific War had come full circle: adaptations grew out of Pearl Harbor and the early carrier battles, were refined in the crucible of the South Pacific, and bore fruit in the central Pacific advance. Forced now to adapt to the USN's countermeasures, the IJN did not have the ability to compete successfully. Its short-war focus had erased its first-mover advantage.

Logistics as Enabler

The story of how USN aviation overcame its early war inferiority would be incomplete without a discussion of fleet logistics. The sustained nature of combat operations necessitated a substantial logistical train.

While U.S. naval leadership had long recognized the importance of logistics in the Pacific, the start of World War II saw the USN as unprepared for supporting the fleet as in other important areas. Between 1925 and 1940, the number of its combatant vessels had doubled, but the number of auxiliaries did not.[112]

These shortcomings in logistics were due more to budgetary austerity than to a lack of vision on the part of the USN. War Plan Orange required speed if the fleet was to rescue the Philippines before it fell to Japan. Behind the quickly-assembled battle force would follow the "fleet train," consisting of oilers, ammunition, cargo, troop transport, and hospital ships.[113] The combatants would need at least one refueling in order to get to the Western Pacific. It was assumed in the mid-1920s that superiority would have been decided quickly thereafter in a Jutland-style action. If the United States was victorious, there would be plenty of time to make up deficiencies in logistics as the enemy fleet would not be a concern. If the United States lost the decisive engagement, the point would be moot anyway.[114] This short-war strategy still required staggering logistical resources; 1920s era planning required the fleet train to exceed 200 vessels, half of which were to be tankers. Budget constraints prevented the USN from acquiring the majority of these vessels, however.[115]

The speed demanded by War Plan Orange, coupled with the lack of available bases, meant that stopping at bases to refuel was not feasible.[116] The fleet would have to refuel at sea. In the interwar years, this was considered a dangerous operation. The vessel to be refueled would have to pull alongside the oiler, maintaining the same speed and course while lines and hoses were passed back and forth in order to transfer the fuel. Even though tests were successfully carried out in 1925 without incident, the USN nonetheless judged this to be a hazardous operation in anything but calm seas.[117] Until the mid-1930s, no plans existed for the use of this technique, because budgetary pressures left fleet commanders and ship captains unwilling to risk damage to the ships or their careers on a seemingly hazardous maneuver.[118] Fleet exercises and operations began to change this view, however. As carriers joined the fleet, so did their increased demand for fuel. Not only did the ships need fuel oil to maneuver, but also the carrier's air wing had its own fuel demands. During fleet exercises, carrier operations were curtailed by the lack of fuel. Also, the Amelia Earhart search and rescue operation, in which carriers had to travel long distances from the West Coast and back with air operations in between, demonstrated that carriers' effectiveness could be drastically undermined if they were short of fuel. Carrier operations during the interwar years influenced the USN to acquire more oilers when fiscally possible and examine methods of refueling at sea more closely.[119]

After Pearl Harbor, the USN suffered from the fiscal hardship of the 1930s, both in general terms and in logistics. An early-war example of logistical difficulties is illustrative of the USN's position:

> After the Japanese bombing of the fleet at Pearl Harbor on 7 December 1941, the three battleships capable of steaming were ordered to the United States for repairs: The *Maryland* and *Tennessee* to Bremerton, the *Pennsylvania* to Hunters Point, San Francisco. The *Colorado* was already undergoing overhaul at Bremerton. When the work was finished, this group assembled on 31 March 1942 at San

> Francisco. There they were joined by the *New Mexico*, *Mississippi*, and *Idaho*, which had been rushed from the Atlantic. Together with a squadron of destroyers which had no tender, this seven-ship force [was] based [at] San Francisco until late in May. The ships were serviced almost entirely from shore facilities. With the exception of targets, target-towing vessels, and planes they were given very little floating service.
>
> On 14 April the force left port with the possibility of being used to assist in stopping the Japanese in their South Pacific drive toward Australia. No train (group of supply vessels) was available, so the ships were crammed with all the fuel, food, and ammunition they could hold. So heavily overloaded were they at the start that they were three to four feet deeper in the water than they were ever meant to be. The third or armored decks were all below the water line; none of the ships could have withstood much damage either above or below water. The Coral Sea action was fought before they could take part in it, the enemy backed off, and the force was not called upon. After staying at sea until their fuel was nearly gone and the fresh provisions exhausted, the ships returned to California at San Pedro.[120]

Only a few months prior to this incident, these ships represented a substantial portion of the USN's combat power. The USN could support small carrier task forces in their pinprick raids against the IJN, but it could not muster the necessary support to get the West Coast battleships into action. The USN still considered these ships valuable fleet units, but it was already making decisions that started to marginalize the battleship.

U.S. industrial might quickly started redressing the fleet's logistical shortcomings. By the end of 1943, it could call upon two large mobile service squadrons, capable of keeping task forces at sea for months, while most navies measured their at-sea time in weeks. From 1940 to 1945, the number of oilers had increased from fourteen to sixty-two, and the number of repair ships from two to twenty-one.[121] Besides the huge quantities of fuel oil and aviation gasoline they carried, the service squadrons also contained mobile dry-dock and repair facilities, enabling combat vessels to return to action faster than if they had to return to the shipyards of the continental United States.[122] In 1940, the Navy had one mobile dry-dock, which was capable of handling a destroyer-sized vessel. By 1945, it had acquired eight more, in addition to three capable of handling battleships, and two for carriers.[123]

The U.S. Navy's Organization, Phase I

Despite its focus on the battleship and resource scarcity before the war, organizational flexibility allowed the USN to put material superiority, which it gradually built through the first two years of the war, to good use. Existing organizational arrangements worked in the USN's favor when facing the IJN's innovative use of carrier air power.

The U.S. Navy's Resources

The resources at an organization's disposal are an important factor in its ability to adapt. In fact, they are a vital condition for adaptation; without resources, little adaptation can take place. As the course of the Pacific War demonstrates, the USN was able to recover from crippling losses, while the IJN was not. The USN had the ability to repair even heavily damaged ships relatively quickly. Those that were sunk were eventually replaced. It is extraordinary that even after losing numerous carriers throughout the first years of the war, U.S. shipyards continued to produce battleships, even while they were launching enough carriers to eventually smother IJN air opposition.

Harder to replace than materiel were trained pilots, and the USN increased its capacity in that area. The USN also struck the right balance between instilling adequate skill in its pilots and getting them to the front in a reasonable amount of time. Spare pilots were scarce at the outset of the war; however, the USN rapidly expanded its training program. Once pilots gained combat experience, the organization made sure that experience was preserved by rotating its veterans out of the front line so that they could train new pilots. The IJN, with its cumbersome pilot training pipeline, could not compete: "A strictly limited number of even the most skillful pilots could not possibly win out over an unlimited number of able pilots."[124] As the war progressed, the number of skilled IJN pilots became even more limited. The tempo of operations did not allow the same rotation as in the USN, and the original IJN cohort virtually disappeared as its members flew until they were killed in action.

As noted above, USN leadership, most notably Commander in Chief Pacific Fleet Admiral Nimitz, realized the difficulties in supporting the far-flung campaigns of the Pacific War. Not only did the USN commit vast resources to building combat vessels and aircraft, but it also produced fleets of logistics ships. These ships supplied the fleet's voracious demand for fuel, food, spare parts, and replacement of both men and materiel. The fleet's service squadrons acted to accentuate U.S. materiel superiority by keeping ships at the front for longer periods of time, and by returning damaged ones to action quicker. In the vast distances of the Pacific, this logistical support was vital. Carrier task forces could conduct sustained operations without worrying about running out of fuel or ammunition. The long march across the Pacific envisioned by War Plan Orange was now feasible.

The Formation of a Dedicated Organizational Subunit

Resources alone cannot explain the USN's success in countering the IJN's innovation. Flexible organization was the necessary condition. Prior to the outbreak of war, the USN had a subunit that did not subscribe to the dominance of battleships within the organization. In the interwar years, the USN was divided as to which was its capital ship: battleships or carriers. The surface warfare community held the upper hand, and the battleship was still dominant. The naval aviation community, however, developed its capability to support the battle line while aspiring to replace it. One of the keys to the USN's success was that it tolerated a subunit within its own organization that had aspirations to be the dominant combat arm of the service. In official doctrine and in practice through the interwar years, naval aviation supported the battle line. In developing its supporting role, however, it also became a combat arm in its own right. Hence, it was in a position to take over the top spot if something happened to the battle fleet. In this way, the USN was organizationally flexible. Because of this, the organization was able to shift its focus from a battleship confrontation that it could no longer win after Pearl Harbor to one based on the fast-carrier task force model. This is all the more remarkable given that the USN was able to do this in the midst of a Great Power war.

The USN's supreme misfortune in suffering the losses at Pearl Harbor also meant that only one subunit, naval aviation, was left to deal with the Japanese innovation. The battle force was still being salvaged at Pearl Harbor, or was stuck out of position on the western coast of the United States while the early decisive battles were being fought. When the battleships started to return to the fleet, both those that were repaired and those of new construction did so as supporting elements. By the time battleships started returning to the fleet in small numbers, the carrier had proven itself as a potent striking platform. Meanwhile, the battleships that had seen action proved vital in fleet defense. The roles of the two ships had reversed. The carrier was now the center of naval operations, and the battleship was the vital auxiliary. Given that the October 1941 strategic estimate predicted that the number of battleships would far outpace the number of carriers even into 1947, the virtual elimination of the battle line from the Pacific Fleet allowed the USN to shift resources to naval aviation in a way not possible in a battleship-heavy fleet.[125] Battleship production continued, but the shipbuilding program put more emphasis on carrier production. Figure 3.1 illustrates how the destruction of the U.S. battle line "cleaned up" the USN organization relative to the adaptation requirement. A requirement that overlaps multiple subunits' missions, or one that does not include any, will make it more difficult for the adapting organization to counter the innovation. Before Pearl Harbor, the USN was divided as to which was its capital ship—the carrier or the battleship. The battleship was still clinging to its dominance within the USN, despite the advances of U.S. carrier aviation. The innovation, IJN improved anti-ship bombing, threatened both carriers and battleships. Because the struggle for dominance as the USN's capital ship remained undecided until Pearl Harbor, the adaptation requirement was competitive until that surprise attack. Once the battleships were out of action, the USN faced a less complicated adaptation requirement (see figure 3.1). The USN had a DOS with which to counter the IJN's innovation, because the carrier and naval aviation were the only tools available. Haphazard affairs as they were, the Battles of Coral Sea and Midway helped solidify the carrier as the center of naval power. Battleships were put out of action *en masse* at Pearl Harbor, but came back to the fleet in small numbers. Battleship doctrine called for concentration of force, and this piecemeal return hampered their employment along traditional lines. Protection

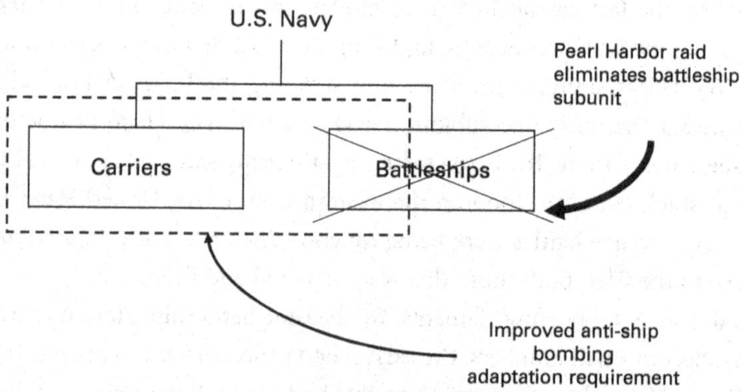

Figure 3.1
The U.S. Navy's adaptation to the Imperial Japanese Navy's innovation.

of the carriers made more sense than forming a battle line for the number of battleships available. The only place where these ships were concentrated was the western coast of the United States, which was far from the battle front. While carriers were at the forefront, earning their place as the USN's capital ship, the battleship was knocked off its pedestal.

Carrier aviation was well positioned to take the battleship's place administratively as well. Its supporting organization, the Bureau of Aeronautics, occupied a co-equal position with the other bureaus within the Department of the Navy. The BuAer was experienced in procuring aircraft and personnel, and it quickly expanded its capabilities both immediately prior to the outbreak of war and after war began. Even though the battleship maintained some of its primacy within the U.S. fleet before Pearl Harbor, carrier aviation was not neglected organizationally, which allowed it to quickly meet the IJN's challenge.

Operationally, the task force concept, in use since 1940, gave the USN the flexibility it needed in constructing its combat forces. It was experienced in putting together mission-oriented force packages before the war began. Carriers, cruisers, destroyers, and battleships had all operated together during fleet exercises in the past. When it came time to use the USN's increased resources, they could be committed in a focused, coordinated way. Even though the primary element of those pre-war task forces, the battle line, was absent after Pearl Harbor, the rest of the fleet units had worked together and could readily form task forces around carriers.

Prior to World War II, the IJN's naval aviation innovation posed a competitive adaptation requirement for the USN. As discussed in Chapter 1, when two subunits can legitimately lay claim to part of the requirement, potentially wasteful bureaucratic infighting can result. While the battleship was nominally dominant in naval doctrine, the carrier was gradually fulfilling more important roles. Even battleship admirals viewed air supremacy in naval war as vital for success. Air power advocates within the USN accepted the supporting role of carrier aviation, but saw the carrier as eventually ascendant over the battleship. The IJN attack at Pearl Harbor turned this competitive requirement into an ideal one for the USN. After December 7, 1941, there could be no real counterproductive internal competition between advocates of battleships and carriers as to which system would dominate. The only real choice was the carrier.

The IJN's innovation in naval aviation was not complete. It made tremendous strides in naval aviation, but it had allowed a disabled enemy to leapfrog its advances. The lingering IJN emphasis on the battleship and a quick war led it to develop an air arm that was ill-equipped to face a recovered USN. While the eventual dominance of the carrier was far from assured prior to the outbreak of war, the IJN raid on Pearl Harbor brutally simplified the USN's choice. The carrier was all that was left with which to meet the IJN's challenge.

While demonstrating the positive effect of a DOS on wartime adaptation, this case study also indicates that adaptation originated from within the USN. The alternative theories of civilian intervention and competition are not as persuasive. Successful adaptation in this case did not originate from civilian intervention. It was the USN leadership that recognized that aviation was the only tool available at the outset of the war. Also, the actions of the Japanese—-i.e., their attack on Pearl Harbor—accomplished half of the organizational task of getting to a DOS by essentially cutting the battle fleet out of the picture. Civilian intervention was not necessary for the USN to adapt.

While naval aviation competed with the battle fleet for influence and resources before the war, the competition theory is likewise not persuasive here. Competition would have meant that there had been a substantial fleet of battleships at the front, attempting to demonstrate that this type of capital ship was still effective at destroying elements of the enemy fleet. This obviously was not the case. Once the battle fleet was knocked out of the war, naval aviation had little competition for the role as center of the fleet.

Innovation, Phase II: "Special Attack"

In response to their worsening strategic situation, the IJN resorted to suicide attacks in late 1944. First appearing at the tail end of the Battle of Leyte Gulf, *kamikazes* made life very difficult for the U.S. Pacific Fleet during the Philippines and Okinawa campaigns. When a *kamikaze* struck a ship, the vessel sank 2.9 percent of the time.[126] While this percentage seems low, it represented a huge improvement in the IJN's tactical effectiveness. At Okinawa, "*kamikazes* sank 27 ships and damaged 164, while [conventional air attacks] sank 1 and damaged 63. A fifth of all *kamikazes* were estimated to have hit a ship—almost ten times the success rate of conventional attacks."[127] The *kamikaze* program turned a large proportion of Japanese aircraft into *de facto* cruise missiles, changing the nature of the threat posed to U.S. ships and dramatically increasing the Japanese naval air arm's effectiveness. Below I discuss the nature of the *kamikazes* and the U.S. adaptation, and then evaluate the USN's organization relative to this new threat.

Japan's decision to adopt *kamikaze* tactics originated from the empire's rapidly deteriorating strategic situation *vis-à-vis* the United States. An imperial council meeting on September 30, 1943, concluded that the empire was overextended. Therefore, Japan needed to identify those areas that were truly strategically vital and contract its defense perimeter accordingly. The council identified the vital areas as the Bonin, Caroline, Mariana, and Sunda Islands, Western New Guinea, and Burma, as well as the sea lanes of communication with them. By contracting the defense perimeter around the Home Islands, the Japanese leadership hoped to give its fleet time to rearm and retrain for the coming U.S. advance. Unfortunately for Japan, however, the tempo of the U.S. advance was too great, and IJN fleet replenishment, rearmament, and training did not progress. Without a respite, the IJN was forced to commit its depleted elements to battle. With the capture of the Marshall Islands and the crippling of Truk, the IJN's main fleet base, in February 1944, vital strategic areas were already vulnerable. From Japan's perspective, "Under these circumstances, it was practically impossible to continue fleet training."[128] The decision to contract the perimeter to better suit Japan's national resources and military capabilities was made too late, however.[129]

The U.S. advance forced Japan to battle before it was ready. When the Marianas were threatened, the IJN had no choice but to commit most of its strength to repelling the U.S. thrust. Furthermore, it felt compelled to commit the bulk of its remaining fleet to the Leyte Gulf action. Both of these battles ended badly for the IJN. Its air arm was decimated at the Marianas Turkey Shoot, and it lost the bulk of its carriers and many of its remaining surface ships at the Leyte Gulf.[130]

The IJN first attempted *kamikaze* tactics in the Philippines in October 1943. These "first attack missions planned and ordered by competent commanders" achieved results that "were unexpectedly good in comparison to the small force committed."[131] At the time, however, the IJN leadership viewed these tactics as tragic, desperate measures forced upon the local commanders. With the worsening war situation, special attack weapons and tactics had been under consideration for some time: "These desperate attack tactics employed by field forces were inevitably reflected in the policies adopted by those concerned with war preparations."[132] In October 1943, the Japanese Naval Ministry and Naval General Staff agreed to produce dedicated *kamikaze* weapons; production began in early 1944, and thousands of special attack weapons were available by the end of that year. These weapons included thousands of specially designed suicide boats and submarines, but the primary weapons were aircraft. The aircraft included conventional fighter, bomber, and training types already in existence. New production types were optimized for ease of production, which meant simplified navigational equipment and other measures commensurate with one-way missions. Pilots needed much less training for *kamikaze* missions than for conventional ones. Less complicated planes and lower training demands provided temporary relief to the IJN's production and training pipelines. When sent on a special attack mission, these aircraft would be fitted with one to three bombs, which increased the attack's destructive power. On the U.S. side, "we had become aware of the Japanese *kamikaze* tactic. It manifested itself sporadically, I guess, as early as early '43, but it was quite obviously a planned and organized chosen tactic by the time of '44."[133]

The *kamikazes* made their first major appearance at the tail end of the Battle of Leyte Gulf. The nature of the attack was immediately obvious to the USN. As the aircraft approached their targets, they were not damaged; they just "came in" and hit the ships. These were not the spontaneous acts of pilots who were about to be shot down anyway. As Admiral John Thatch stated, "It was actually a guided missile before we had any such things as guided missiles."[134] While the *kamikazes* arrived too late to affect the course of that battle, they inflicted serious damage in their opening salvo, sinking the escort carrier *St. Lo*.[135] In order to avoid anti-aircraft gunnery and the U.S. CAP, *kamikaze* flights would use terrain to mask their approaches by hugging nearby island hills and valleys. They took off from multiple dispersed bases, and approached their targets from multiple directions. In order to hide their approach, the *kamikazes* often tried to closely follow returning U.S. air strikes back to the fleet, negating the IFF system by hiding in the U.S. aircraft radar return.[136] As they made their attack runs, the U.S. CAP would be divided and confused, and the target's gunnery would be split, decreasing its effectiveness.[137] The *kamikazes* did not attack their targets alone. They were often accompanied by conventional fighters and torpedo planes, which would distract the U.S. defenses and help the special attack aircraft get through.[138]

This innovation forced a change in USN strategy. When the IJN had been using conventional naval aviation, the USN's ship gunnery could turn back or deter their strike aircraft before their attack runs. The early-war example of the USS *North Carolina* is illustrative. The battleship used its secondary battery of 5-inch guns *en masse* to engage IJN aircraft at long range. The ship's commanding officer noted that "barrage fire was highly effective in dispersing attacks. The barrages were very heavy and the enemy planes refused to come through the bursts. It is doubtful if any planes were shot down by barrages laid by this ship, but many attacks were dispersed."[139]

The special attack missions, on the other hand, were a nasty shock to the U.S. fleet. The main reason for this was that they were so hard to deter, and doing so was complicated by the tempo of these operations. While conventional air raids could be deterred, *kamikazes* kept going, in part because they did not need to survive to fight another day. The number of guns a ship mounted did not matter in this regard; U.S. ships mounted more guns than they did early in the war, but the increased armament did not translate into deterrence because there was no way to deter a *kamikaze*. The targeted ships would have to wipe out an approaching wave of aircraft in order to avoid being hit. Furthermore, gunners could never be sure that a particular *kamikaze* aircraft would not hit its target unless it was completely destroyed: "With a *kamikaze*, the plane you had to shoot down was one who was going to get as close to you as he could get, preferably crashing on the deck, and if you hadn't killed him or torn him to shreds, you were interested in continuing to try and do that until the last instant before he landed on board."[140]

Gaining extraordinary momentum in their final steep dives, *kamikaze* planes proved capable of hitting home without their wings, whether stripped off by the dive or shot off by the U.S. gunners.[141] *Kamikaze* raids were frequent, and they proved resilient to counterattacks on their bases.[142] This inexorable and alien form of attack induced a new level of stress on the USN: "One gunner had the harrowing experience of being knocked down by the bloody upper half of a Japanese pilot's torso as it hurtled through the air after the plane exploded."[143]

The appearance of the *kamikazes* during the Philippines campaign not only affected the USN's fleet, but also operations ashore. As the primary source for air support of the ground forces, the USN's aircraft carriers played a vital role. With this rough treatment at the hands of the *kamikazes*, the fleet suffered and could not provide the ground forces with as much air support as it needed. The bulk of U.S. carrier strength, based on the fast fleet carriers, had to withdraw temporarily to recover. The ground campaign slowed as a result.[144]

The USN relied upon its previous adaptations to IJN carrier aviation to meet the *kamikaze* threat. Beyond that, its primary method of adaptation was to use its advantage in numbers. It made important adjustments in its air defense methods, but accomplished the lion's share of the adaptation by attempting to smother its adversary with newly constructed ships and aircraft.

When the fast carriers returned to the Philippine theater, they came with increased fighter complements—the previous 54 fighters per carrier group had been increased to 73—and a new plan. The fleet would use its massive superiority in aircraft to lay an "air blanket" over potential or suspected *kamikaze* bases. In effect, the USN carrier groups had enough planes to strike each base twice a day, as well as maintaining a CAP over it for much of the daylight hours. In theory, the *kamikaze* bases would not be able to recover from one raid before the next, and if they did, the launching planes would be intercepted as soon as they became airborne. The air blanket proved to be a partial success, but there were just too many bases to cover. Also, many *kamikaze* bases were heavily defended by anti-aircraft artillery, which began to take a heavy toll on USN carrier air groups.[145]

Tactically, USN aviation further adapted by using its own aircraft in war games to simulate *kamikaze* methods, which aided in devising countermeasures. These exercises were codenamed "Moose Trap." The naval aviators emulated *kamikaze* tactics in the exercises, namely by approaching their targets from different altitudes (i.e., one plane high, another low), and taking advantage of blind spots in radar coverage. In addition, they imitated the IJN tactic of closely following U.S. strike aircraft back to their ships to negate IFF.[146] The Moose Trap exercises were instrumental in developing countermeasures for protecting the USN's fleet.

The resulting system used radar-equipped destroyers and large numbers of fighters on CAP. The destroyer pickets would have fighter direction and command of eight CAP fighters, and they would sail fifty miles out from the carrier task force in order to increase warning time. Also, returning U.S. strike aircraft would orbit over a predetermined destroyer picket while its CAP would ensure that *kamikazes* and other aircraft were not hiding among U.S. aircraft in order to counter the IJN method of negating the U.S. radar and IFF systems. The fleet also delineated "fair game" areas in which any air contact would be engaged. The *kamikaze* threat was so dire that even friendly aircraft that strayed into these areas risked being shot down.[147] It was too risky to give unidentified blips on the radar the benefit of the doubt, because once *kamikazes* were close to the invasion fleet, they were very hard to stop. This defense system, developed almost entirely by the USN's fleet in the combat zone, was later used during the Okinawa operation.

The Philippines *kamikaze* campaign ran out of planes in January 1945. It had sunk 24 USN ships, damaged 67, and killed 800 sailors and wounded 1,400. This campaign was proportionally more devastating that that off Okinawa later in 1945.[148] IJN naval aviation at the Marianas was a shell of its former self. The *kamikaze* innovation helped it recover its striking power almost overnight.

During the capture of Iwo Jima early in 1945, a *kamikaze* strike succeeded in sinking an escort carrier and heavily damaged a large fleet carrier, the *Saratoga*. Having been misidentified as "friendlies," the approaching IJN planes snuck through the task force's defenses. Only one aircraft actually struck the

Saratoga, but the carrier's decks were full of armed and fueled planes. The hit made a thirty-foot hole in the flight deck and caused a massive fire, putting the ship out of action for months.[149]

The longest-running and most famous *kamikaze* campaign was during the Allied operation on Okinawa. On the doorstep to the Home Islands, the Japanese had little choice but to attempt a maximum effort to at least slow down the U.S. advance. They assembled thousands of aircraft for special attack duties, intending to cripple the U.S. fleet while it sat immobile off Okinawa supporting the invasion. Unlike in previous amphibious operations, the USN fleet had decided to stay in close proximity to the shore in order to better support the land campaign. The tradeoff of this decision, however, was that the USN forces lost their mobility. Tied to the invasion beaches, the USN's ships at Okinawa made easy targets for the *kamikazes*.[150] There were no other major operations during this time, so the IJN could commit more than 7,300 planes to the effort. Also, these planes were based in close proximity to the U.S. fleet, both on Okinawa and the Home Islands. The curriculum at the U.S. Naval War College identified them as "the major threat to our Naval Forces."[151]

The array of *kamikaze* weapons at Okinawa was formidable. Not only did it consist of a variety of conventional fighter, bomber, and torpedo aircraft, but it also included the "Okha," or "cherry blossom." Dubbed the "baka" ("fool") bomb by the Allies, this was a missile manned by a human pilot. After launch from a Betty twin-engine light bomber, the rocket-propelled weapon reached speeds of 500 miles per hour and delivered a 1,200-pound warhead. One hit was capable of sinking a destroyer outright. Its high speed meant that it was hard for the pilot to make even minor course corrections, however, and many overshot their targets.[152]

As the U.S. fleet approached the shores of Okinawa and supported the landings, it set up a ring of radar pickets to provide early warning of air attack. These consisted of widely-dispersed destroyers equipped with radar. They sailed at sixteen stations around the fleet, fifty miles apart.[153] This was the first operation of the war in which the pickets had heavy CAP support and full control over it through fighter direction facilities.[154] As the first line of fleet defense, they were the first to fall victim to the *kamikazes*: "The Japanese concentrated on the outlying picket vessels, believing that once they were knocked out the rest of the fleet would be easy targets."[155] The result was that the destroyers suffered heavily at the hands of the *kamikazes*.

As the Okinawa operation progressed, the picket system evolved. While the picket destroyers started the operation sailing alone, the threat prompted fleet commanders to strengthen each position with gunboats to provide extra firepower and render assistance to stricken vessels.[156] In the final month of the campaign, each picket station consisted of three destroyers, four landing ships serving as gunboats, and extra gunboats when heavy attacks were expected. Heavy CAP was stationed in the vicinity of the picket stations.[157]

During forty consecutive days of *kamikaze* attacks, both large and small, the ships of the USN's invasion fleet were at "condition one," which meant that at least two-thirds of the crew were at their battle stations at all times, or "general quarters," which meant that all of the crew were at their battle stations and all watertight hatches were secured. The ships' boilers were also ready to give full power at a moment's notice. The consequences of this prolonged state of alert should not be underestimated. The ships' guns were constantly manned, and the temperature in the boiler rooms topped 100 degrees Fahrenheit. The resulting combat stress and fatigue posed major problems for the fleet.[158]

The *kamikazes* were able to inflict severe damage on the USN picket line, but were not very effective in getting past those outer defenses to strike decisive targets. The ordeal of the pickets is borne out by the numbers. Of nineteen destroyers originally fitted with fighter direction facilities, four were sunk and eight seriously damaged. The USN sent fourteen replacements; of those, one was sunk and five were seriously damaged. Therefore, of the thirty-three total destroyers equipped to direct the CAP, five had been sunk and thirteen seriously damaged.[159]

The effort needed to achieve this result, however, was at times excessive from the standpoint of IJN effectiveness. The example of the USS *Laffey* is illustrative. While on picket duty, this ship was attacked by twenty-two IJN planes. It suffered four bomb hits and six aircraft hits, but remained afloat. It shot down nine of those planes before they could do any damage. By the end of the action, though, the ship had only two small-caliber guns still in operation.[160] The first several hits by IJN bombs and *kamikazes* undoubtedly caused a "mission kill," making the *Laffey* useless as a picket ship. It is hard to believe that the ship's electronics and power survived the shock and blast caused by those explosions. In trying to sink this one picket, however, the IJN inadvertently allowed the crippled ship to continue to fulfill its roll in protecting the fleet.

During the Okinawa operation as a whole, "these ships and their supporting craft absorbed the brunt of 846 raids from approximately 3,000 planes."[161] The *kamikazes* sank 21 ships, and damaged 217; 20 of those ships were hit multiple times. The USN detected an overall drop in *kamikaze* performance compared to the earlier campaign. Most attacks were direct, low-flying, and executed with little maneuvering. Many of the planes were shot down by "relatively unarmed ships." The *kamikazes* seemed to aim for small, isolated ships that didn't have sufficient firepower to repel them; however, the escort carriers supporting the land campaign were relatively untouched.[162] By hitting the wrong targets, "the *kamikazes* had failed."[163]

Radar and the proximity fuse proved to be the most effective technological counters to the *kamikaze*. Radar was the "heart and soul" of air defense, while the 5-inch proximity fuse was the biggest killer of incoming aircraft.[164] With the U.S. fleet getting battered at Okinawa, the development of countermeasures took on a fevered pace. The USN's leadership recognized that the effort needed to be more coordinated and centralized. The commander of the Pacific Fleet and the Chief of Naval Operations, Admirals Nimitz and King respectively, created an organization that came to be known as Composite Task Force 69 (CTF 69), or the Operational Test and Evaluation Force.[165] Led by Vice Admiral Willis Lee, a gunnery expert and veteran combat leader, its task was to develop weapons and tactics to combat the *kamikaze* problem. The USN gave this organization command of an old battleship, the gunnery training vessel *Wyoming*, and three squadrons of aircraft manned by veteran pilots. Also, it seconded newly launched ships to this command in order to evaluate their sensors and weapons and further refine anti-aircraft methods.[166] The efforts of the task force were "difficult to document because we were under such a press of urgency that most of our decisions were made sort of by the seat of the pants on the spot, given by telephone to Washington and action was taken accordingly."[167] With its veteran personnel and mixture of equipment, the task force "could reproduce any set of hypothetical tactics that you wanted."[168] The efforts of CTF 69 came too late for the Okinawa operation, but helped the Pacific Fleet better cope with the *kamikaze* threat in the months leading up to Japan's surrender. Had the Allies proceeded with the planned invasion of the Home Islands, they faced a potential *kamikaze* opposition of more than 6,000 planes.[169] It is impossible to evaluate how this *kamikaze* force would have fared against an invasion force equipped with the new techniques developed by CTF 69, but given the declining effectiveness of the *kamikaze* over time at Okinawa, it is reasonable to assume two things. First, the invasion of Japan proper would have been a much larger operation, meaning the targets would have been more plentiful and the casualties inflicted by *kamikazes* greater. Second, assuming the trend from the Philippines to Okinawa would continue, the proportion of the casualties inflicted by *kamikazes* would have been smaller than in previous operations.

A number of factors contributed to the steady reduction of the tempo of *kamikaze* attacks. The U.S. fleet increased its CAP and picket ships and improved their fighting techniques. This, coupled with the one-way nature of *kamikaze* missions, combined to inflict heavy losses on the IJN. The latter also suffered from volunteer and fuel shortages as the campaign went on. Finally, the United States used its massive carrier air force, along with some U.S. Air Force B-29 missions, to bomb the *kamikaze* airfields on an almost round-the-clock basis.[170]

Strategically, the *kamikazes* had failed. They were unable to stop the U.S. advance through the Philippines or the capture of Okinawa. They did, however, increase IJN aviation's operational effectiveness almost overnight. I attribute this result to both the nature of the innovation and the USN's organization.

The U.S. Navy's Organization, Phase II

Because of its earlier adaptation to innovation in IJN carrier warfare, the USN was well-placed to adapt to the *kamikaze*. It used a combination of its unparalleled access to resources and air defense doctrine developed over the previous two years to combat the threat. The organization also quickly recognized the innovation, developed a response, and put that response into effect. This case, however, demonstrates that an innovation can pay telling operational dividends even against an opponent with both good organization and sufficient resources. This becomes clear as we look at USN resources and organizational arrangement.

Resources

The USN adapted to the *kamikaze* primarily by increasing the numbers of its ships and aircraft. Hence, the heart of its capacity to adapt was its access to resources. As the narrative above makes clear, relatively minor tactical adjustments in fleet air defense were accompanied by massive outlays of resources. The USN could call upon dozens of aircraft carriers and their air groups. It also possessed hundreds of supporting ships that contributed to the air battle by providing early warning, fighter direction, and anti-aircraft gunnery.

Technology also played an important role in the USN adaptation. Radar and IFF gave U.S. ships early warning of attacks. The proximity fuse made the heavy anti-aircraft guns of the fleet much more accurate. Fighter direction facilities on picket ships made intercepting *kamikaze* flights more efficient. Without these technologies, the USN would have suffered a great deal more than it did. These developments were instrumental in forming a layered defense around the fleet. Fighters formed the outer layer; the picket ships both directed the fighters in their task and formed the next layer of defense with their guns; the ships of the carrier task force made up yet another layer with their guns. Ideally, by the time an attack reached the carriers themselves, it was heavily depleted or dispersed.

As important as these technological innovations were, though, they were still somewhat inadequate against the *kamikaze* threat. Optimized for use against conventional air attack, the USN's radar and guns did not erect an air-tight seal around the fleet. Against a conventional attack, this was theoretically possible: shoot down enough aircraft in a raid, and the rest would be deterred from driving the attack home. A *kamikaze* raid was qualitatively different because, first, the attackers could not be deterred by casualties because they did not plan on surviving, and second, only one hit on a major fleet unit (e.g., a carrier or amphibious warfare ship) could be catastrophic. The USN was essentially confronted with the anti-ship cruise missile threat years before such a thing existed. Even the most advanced technology of the day was inadequate against such a threat.

The Formation of a Dedicated Organizational Subunit

The above narrative reveals that the threat the USN faced from *kamikazes* was similar to that posed by conventional IJN carrier aviation. It was still largely an air threat, requiring the same kinds of adaptations. Therefore, the second USN adaptation depended a great deal on the first. The attacking *kamikazes* consisted of similar or identical aircraft to those previously used by the IJN. Their targets were the same: the U.S. carriers and the landing ships they supported. The *kamikaze* also challenged the same elements of the U.S. fleet as their conventional brethren: radar and IFF, CAP, air defense doctrine, and the USN's ships' ability to absorb damage. Given that the USN had forged its carrier fleet through more than two years of combat, it was even better placed to face the *kamikaze* than it was to face the initial IJN innovation in carrier warfare. While the battleship reappeared in the U.S. fleet in the form of new vessels and the repaired units from Pearl Harbor, by the time of the *kamikaze*, it was an integral part of the carrier task force, not a competitor for operational dominance of naval operations.

The Pacific Fleet not only recognized the *kamikaze* threat, but also used its operational units to analyze and predict future *kamikaze* tactics. It modified its operations accordingly, so that the *kamikazes* were less able to exploit its weaknesses. In addition, it changed the components of its air groups to better deal with the threat. The lessons learned in the retaking of the Philippines were quickly put to use in the Okinawa campaign.

Given the tempo of operations, however, it was not possible to adapt to the threat completely. USN leadership established CTF 69 very quickly, but not in time to influence the outcome at Okinawa. As noted above, technology that proved its worth in turning back conventional air attacks was simply not as effective in fighting the *kamikaze*. The task force was developing new technology and methods to address these weaknesses as the war ended. It is reasonable to assume that its efforts would have borne fruit during a final invasion of Japan, but one can only pose hypothetical notions in that regard.

The USN's adaptation to the *kamikazes* came from within the organization, from the top down. The alternative explanations of civilian intervention and competition carry little weight. There is little evidence that civilians had any effect in pushing the USN toward adapting to the *kamikaze* threat. The damage and casualties suffered by U.S. ships in the Philippines and Okinawa was enough to spur the USN into action.

Likewise, there is little evidence of competition between subunits producing the adaptation. The USN was essentially on its own, and within the USN, there was little question that carrier aviation would bear the burden of fending off the *kamikazes*. The surface fleet played a large role in air defense, but it did not try to supplant aviation in its role as the first line of defense. Both elements worked together closely—both experimentally and operationally—to meet the threat.

Conclusion

Against improved IJN anti-shipping capability, the USN leveraged its good organization to perfect its adaptation, the fast carrier task force, which was superior to the original innovation in almost every way. When the USN was short of resources at the beginning of the war, it used its strong organization (relative to the innovation) and centralization to aid its holding action against the IJN and develop its countermeasures. By the time it gained resource superiority, the USN's mode of operations had improved to such a point that the combination of superior methods and crushing numbers was nearly unstoppable by the IJN.

The USN that adapted to the *kamikaze* threat had the same ideal organization that allowed it to adapt to the IJN's air aviation innovation. By late 1944, it possessed dozens of carriers and their air groups, and it used methods honed and tested by years of combat. When the *kamikaze* appeared, the USN adapted quickly with all the resources at its disposal. The IJN innovation was a strategic failure due in part to those adaptations. The IJN, however, recovered its ability to sink ships for a time. After losing hundreds of aircraft in the Marianas for virtually no gain, it was able to use this new mode of operation to inflict serious damage on the USN. No air defense can be leak-proof, and the *kamikaze* exploited that fact. This case demonstrates that some innovations can be so groundbreaking as to temporarily overwhelm even the strongest organization.

4

The U.S. Air Force and Its Adaptation to North Vietnamese Air Defenses

THE U.S. ARMY was not the only military organization to face a new and challenging enemy during the Vietnam War. The U.S. Air Force (USAF), in carrying out the policies of the Lyndon Johnson and Richard Nixon administrations, faced an enemy that used a new mode of operation. The USAF had spent the years between the Korean and Vietnam Wars focusing on what it perceived as the most likely opponent it would face in the future: the Soviet Union. A senior officer corps that dogmatically believed in the doctrine of strategic bombing and carried it out in World War II led the USAF's organization. The Soviet threat, policy guidance from the civilian leadership, and internal bureaucratic culture all steered the USAF toward optimizing for delivery of nuclear weapons. Within the USAF, Strategic Air Command (SAC), the organization that controlled the U.S. deterrent forces, became dominant over Tactical Air Command (TAC). The latter's mission was to accomplish more conventional missions, such as air superiority and close air support for the U.S. Army. So great was the USAF's focus on nuclear weapons delivery, however, that TAC transformed itself into a "mini SAC" by the time of the Vietnam War.

When charged with deploying the various weapons of war on North Vietnam, it was TAC that bore the larger burden. As Operation Rolling Thunder commenced in 1965, TAC went into combat with fighters and bombers designed for "tactical" nuclear warfare in central Europe. What TAC faced in Southeast Asia was an integrated air defense (IAD) that used a combination of high technology and "guerilla warfare in the sky." The North Vietnamese had at their disposal imported Soviet SA-2 Guideline surface-to-air missiles (SAMs), supported by sophisticated surveillance and guidance radars. Also, USAF bombers faced large numbers of anti-aircraft artillery (AAA). The North Vietnamese used their limited number of SAMs not only to shoot down U.S. aircraft, but also as bait in deadly games of cat and mouse. When the United States tried to neutralize the SAM threat by attacking the Vietnamese bases, they often found either empty emplacements or dozens of AAA weapons.

The USAF, being largely unprepared for this type of threat, had to adapt in the midst of its operations, making this a case of wartime innovation. The simple counter of attacking the SAM emplacements directly was a disaster. The first U.S. efforts to punish the North Vietnamese for using SA-2s were unsuccessful and merely resulted in more downed aircraft. Ultimately, the USAF used sophisticated electronics for SAM detection and radar jamming, combined with specially tailored aircraft known as "Wild Weasels" and trained pilots to neutralize the North Vietnamese defenses. These aircraft carried special anti-radiation missiles (ARMs) in addition to more conventional munitions. This combination steadily whittled away at the effectiveness of North Vietnam's IAD. No adaptation is 100 percent effective, but North Vietnam found itself using more and more of its limited resources to bring down fewer USAF aircraft.

In this chapter, I will first define the scope of this inquiry, which is precisely defined in order to illustrate this study's utility in discovering the effects of wartime innovation. Also, in this case, the question of evaluating the effectiveness of the innovation and adaptations is a sticky one. For the United States, this was a limited war, so calculations of effectiveness must be adjusted accordingly. Second, I will discuss in detail the North Vietnamese Navy (NVN) innovation—its IAD network, which consisted of AAA, SAMs, and radar that were used to take advantage of the NVN's limited numbers and the U.S. coercive strategy. Third, as in the other cases in this volume, I will outline the strategic setting, specifically how the USAF was structured to fight as it entered the Vietnam War. As the previous cases have demonstrated, the nature of the adapting organization is very significant in how "innovative" a new mode of operation is. In this case, the USAF was so focused on its nuclear delivery mission that it neglected all of its "conventional" responsibilities. This could only help in making the NVN's mode of operation more novel from the USAF perspective, even though the latter had faced IAD before.

Fourth, the USAF adaptations are examined in full. The United States used a combination of technology and organization to overcome the IAD of the North Vietnamese. The technology came remarkably quickly, and the organization changed in a more gradual manner. The most important element, however—the training of a special cadre of officers to combat the SAM—was in place relatively early. Taken together, the NVN innovation and the USAF organization initially produced a "shunned" adaptation requirement. In other words, multiple subunits of the USAF, in this case SAC and TAC, had the need to address enemy IAD in the conventional (non-nuclear) environment, but neither did so. SAC understandably oriented itself toward penetration of Soviet air defenses in a general nuclear war scenario. TAC also assumed that its primary opponent was the Soviet Union, but reasoned that it could wholly avoid Soviet IAD and thus did not develop the capability to address it head-on. The organization then formed a top-level review panel (TLRP) to generate its response.

Fifth, I will evaluate the extent of the success of the USAF adaptations. The ability of the USAF to organize will then be analyzed, along with the viability of the alternative explanations of the source of adaptation. The interaction between TAC and NVN air defenses draws a detailed picture of the USAF's ability to organize to meet the threat. Taken together, this evidence will demonstrate that, like the British Royal Navy (BRN) in World War I, the USAF had many assets in terms of resources. It could draw upon aircraft, personnel, and technical expertise when it needed to. Its organizational arrangement in terms of the adaptation requirement imposed by the air defenses was not ideal, however. North Vietnam's IAD network posed a "shunned" requirement for the USAF, as neither SAC nor TAC could directly address it. The structure of the USAF immediately prior to the conflict left the primary subunit tasked with executing Rolling Thunder unprepared for the threat that it would face. It is also the basis of the resource scarcity that the USAF experienced: electronic countermeasures and specialized aircraft were slow to arrive in sufficient numbers, in part because TAC had not placed a high priority on developing these systems.

The USAF's effective adaptation started with a TLRP. As outlined in Chapter 1, this was a temporary measure, and not as ideal as the creation of a permanent dedicated organizational subunit (DOS), but it improved the efficiency of the USAF adaptation effort by developing a multifaceted solution to the innovation problem.

Scope of Analysis

For this study, the NVN innovation consisted of SAMs and various calibers of AAA working in concert, all aided by radar. I will go into greater detail below about the methods that the NVN used to make this combination effective. At this point, however, it is important to acknowledge that there is a third component to the North Vietnamese air defense network that is not included in this analysis because it would be of limited utility for this study. The NVN use of Soviet-built MiG aircraft was significant. The MiGs posed a hazard to U.S. strike aircraft and gave the North Vietnamese important symbolic victories when they prevailed in air-to-air combat.

The previous cases of successful adaptation in this volume demonstrate that creation of an organizational subunit whose responsibilities encompass the entirety of the new threat is instrumental to an effective response; this case, however, is more complicated. Combating AAA and SAMs on the one hand and MiGs in air combat maneuvering (ACM) on the other are two very different propositions. Today's multi-role aircraft may be able to combine these two missions, but they are still not missions that are easily or efficiently accomplished at the same time by the same aircraft. Furthermore, in the 1960s, the ability of aircraft to accomplish missions as diverse as air-to-air combat and suppression of enemy air defenses (SEAD) was extremely rudimentary at best. Therefore, a single subunit that was built to meet all three aspects of the NVN air defense threat (AAA, SAM, MiG) would have made little sense. The air threat required very different technology and flying skills than those for the SEAD mission.

In addition, there was a relatively simple way for the USAF to address the MiG threat without needing much organizational change. The MiGs flew from a limited number of fixed bases, which were easily neutralized if the United States so chose. A concerted bombing effort by large numbers of U.S. aircraft undoubtedly would have accomplished the task fairly quickly, as there were few other places the MiGs could go. The main reason for the continued effectiveness of the MiGs lay with the political decision not to bomb the airfields. In planning the air campaign that would become Rolling Thunder, the U.S. Joint Chiefs of Staff placed a high priority on targeting airfields, but the rules of engagement (ROE) that dictated the behavior of the USAF prohibited U.S. aircraft from engaging the MiGs at their bases, even in hot pursuit.[1] The fact that North Vietnamese MiGs simply moved to Chinese bases after U.S. forces were free to bomb those in Vietnam does not negate the fact that the MiG problem can be viewed as a political problem first and foremost.[2] Therefore, while it is impossible to entirely omit any mention of the MiGs, an in-depth treatment of the MiG threat would add little to this inquiry.

I have focused my examination on USAF efforts aimed at force preservation. In a more conventional war, an analyst might consider both losses in aircraft *and* the degree of damage inflicted upon the enemy—"measures of effectiveness" (MOE) in modern military parlance—in evaluating the effectiveness of a bombing campaign. The largely political goals of the U.S. air campaigns over North Vietnam make that approach dubious in this case, though. The primary focus of this study in terms of phases of the Vietnam War will be on Operation Rolling Thunder, from 1965–1968. The Linebacker operations of 1972 will also be examined. These two periods offer interesting contrasts because of their different levels of operational intensity and their overall strategic goals. Many credit relatively unfettered use of air power for the success of Linebacker I and II. One of Rolling Thunder's many faults, especially according to those in USAF circles, was its halting, limited nature. Even an analyst that does not subscribe to air power theory is hard pressed to find military utility in the Johnson administration's handling of the campaign. As Robert Pape notes, however, while both Rolling Thunder and Linebacker were intended to coerce the North Vietnamese, the U.S. policy goals for each of these two campaigns were very different. Whereas Rolling Thunder was designed to force the North Vietnamese to give up their designs on the South entirely, Linebacker's purpose was to blunt their conventional invasion of the South and force them to the negotiating table.

In hindsight, Rolling Thunder's goal was unattainable: unification was a permanent interest that North Vietnam would never relinquish. Also, political considerations were the primary driver behind the tempo of this operation and the times at which targets were struck, even though the target list itself was generated by military requirements. Applying measures of military effectiveness to an essentially political series of acts would not make much sense. The goals of Linebacker were more limited and realistic. North Vietnam's military effort was vulnerable to the heavy use of U.S. air power, as I discuss below. Therefore, measuring the effectiveness of Linebacker in terms of targets destroyed would be less problematic. The problem comes in comparing the effectiveness of Linebacker with that of Rolling Thunder. In order to make a valid within-case comparison between the two operations, we must measure both by the same yardstick. In both cases, North Vietnam's innovation and the consequent USAF adaptations were focused on inflicting or preventing aircraft losses, respectively. Thus, it is in terms of aircraft shot down that I will evaluate the innovation-adaptation relationship. This approach controls for the limited nature of Rolling Thunder, the different operational tempos of the two major air campaigns of the war, and their different strategic goals.

This case also demonstrates that successful adaptation need not be simply a function of eventual victory or defeat, justifying my focus on operational (not strategic) results. As this chapter will show, the USAF was relatively successful in reducing its losses to NVN air defenses—but this success occurred within the context of losing a war. This helps, in part, to control for the fact that in my other case studies, the majority of the adapters were the eventual victors in the wider conflicts.

The Innovation: Surface-to-Air Missiles, Anti-Aircraft Artillery, and Radar

The NVN innovation was the use of SAM, AAA, and radar in an IAD network. This innovation was itself a reaction to the U.S. bombing campaign, as the necessary equipment was unavailable until after this campaign started, and it caught the USAF unprepared. It had faced IAD networks before, during both World War II—especially over Germany—and the Korean War. North Vietnam, however, used both new technology and novel tactics that made this challenge unique.

The NVN strategy was to place its limited air defense assets around the targets most likely to be the subject of U.S. attack. At first, these weapons consisted of a few AAA guns, fire control radars, and surveillance radars. As the NVN imported large quantities of guns and radars, as well as SAMs, from the Soviet Union, its existing defenses were reinforced and additional areas could be defended. The pattern of expansion that followed was that the targets around Hanoi and Haiphong were defended first, and then defenses expanded southward as equipment levels increased. This was a target-centered defense; rather than attempt to defend against all airborne incursions, the NVN concentrated its efforts on making it costly for the USAF to accomplish its mission. At the same time, the defenses of individual targets were linked together so that the NVN could use sensors in multiple areas to assist in blunting U.S. attacks. There was an obvious power disparity between the United States and North Vietnam. Therefore, as Earl Tilford states, "the key to thwarting the American aerial assault lay in obviating the strengths of the U.S. air forces in jujitsu fashion. Instead of struggling with the Americans for air superiority, the North Vietnamese opted for a lower level of strategy, one best termed *air deniability*. . . . [It was] a lower, more basic form of warfare constituting, in its essence, a people's war in the air."[3]

The NVN used a wide range of Soviet-developed equipment in pursuit of this strategy, and it used them in a very coordinated fashion, making their air defense network truly "integrated." Tilford notes that "the SAMs posed the kind of threat that Air Force technicians could, and did, overcome quickly."[4] These weapons, however, forced USAF planes to fly in ways that made them most vulnerable to AAA:

> Although 125 SA-2 missiles were observed in flight (presumably, many others were not observed), they downed only 5 USAF . . . and 5

> Navy aircraft. Much of the effectiveness of these Soviet-built missiles had been nullified by their relatively poor guidance system, U.S. electronic countermeasures, and the evasive tactics of U.S. pilots. On the other hand, they forced pilots to fly in lower to hit targets, making them more vulnerable to ground fire and thus indirectly increasing U.S. losses.[5]

The NVN's IAD innovation was itself a reaction to the escalating U.S. bombing campaign. In July 1964, the NVN possessed only 50 anti-aircraft guns, 24 air defense radars, and no SAMs. By April 1965, the United States had detected the first SAM site under construction. The sites were identical to those photographed by aerial reconnaissance in the Soviet Union: a radar van surrounded by missile launchers in a "Star of David" arrangement. During the summer of 1965, the NVN increased the numbers of AAA to 3,000, introduced 100-millimeter guns to its arsenal, and fired its first SAM at a U.S. aircraft during the conflict. In December 1965, the NVN's radar operators started to control their emissions in response to U.S. countermeasures, namely ARMs. By August 1966, the USAF considered the NVN to have completed a "sophisticated air defense system," with 4,400 AAA guns, an estimated 20 to 25 SAM battalions, and 271 air defense radars. In December 1966, it had built further upon this system, and the USAF faced 6,398 AAA guns, and a full 25 SAM battalions rotating among 151 sites.[6]

The most famous element of the NVN air defense network was the "Dvina" SAM, NATO designation SA-2 "Guideline."[7] During the Vietnam War, these missiles were organized into battalions. SAM battalions consisted of eighteen missiles, one for each launcher and a dozen spares. Supporting these missiles were acquisition and guidance radars, electrical generators, and a fire control computer. Approximately twenty-five vehicles transported the entire unit. Each battalion could rotate between prepared sites on relatively short notice. The United States first encountered the Dvina when this missile downed Francis Gary Powers's U-2 reconnaissance aircraft over the Soviet Union in 1960. In 1962, an SA-2 shot down a U-2 over China, killing its Chinese nationalist pilot. That same year, an SA-2 was used to down a SAC U-2 over Cuba during the Cuban Missile Crisis in October. The guidance radar that accompanied the missile was code-named "Fan Song" by NATO, and included altitude, azimuth, and missile guidance radars. The most common acquisition radar accompanying a SAM battalion was code-named "Spoon Rest"; it had a range of about 100 nautical miles.

The SA-2 operated in the following manner. Once the Spoon Rest acquired a target, the information was passed on to the Fan Song. The latter was a "track while scan" set, meaning that it could scan for new targets while it tracked the current one, although it could not engage more than one target at a time. The Fan Song had two trough-shaped radar antennas, one of which scanned horizontally, the other vertically. The former provided azimuth and range data, while the latter provided altitude data and verified the range. The process of acquisition, target lock-on, and missile launch took about 75 seconds. Once launched, the same radars guided the missile to the target. A guidance beacon at the base of the missile provided Fan Song with the position of the missile, and the computer would calculate an intercept point based upon the relative position of the missile and the target. The missile was command-guided to the intercept point. Once within 1,000 feet of the intercept point or 23 seconds after launch, a command signal would arm the missile's radar proximity fuse. If the target eluded the missile, it would automatically detonate between 57 and 63 seconds after launch.

The missile itself could reach speeds of Mach 4 (four times the speed of sound, which varies slightly depending on atmospheric conditions). It measured 34.5 feet long on the launching unit and 26.5 feet after its booster had separated, giving it a distinct "flying telephone pole" shape. Its 420-pound warhead gave the weapon a 150–200 foot blast radius. It had a 25-nautical-mile slant range, and could engage targets at a maximum altitude of 60,000 feet. Conversely, it was ineffective below 3,000 feet, because Fan Song could not track targets below this altitude. Additionally, the SA-2 had a "dead zone" of about five nautical miles, as this was the distance required for the warhead's fuse to arm. Later on in the conflict, the Soviets developed an optical tracking system for the missile. While the new tracking device, which was mounted on the Fan Song unit, required good visibility for proper operation, it was essentially immune to electronic jamming and only required the electronic signals for the uplink and downlink between the guidance system and the missile itself. This allowed the NVN to launch missiles without alerting U.S. pilots and to use one Fan Song to distract the target while an optically guided SA-2 approached from an unexpected direction. Furthermore, it allowed the Fan Song to engage targets as low as 1,000 feet. This upgrade was put into place in 1968 and did not figure prominently until the Linebacker operations.

Anti-aircraft guns of almost every description complemented the SA-2, and the NVN used them to great effect. These weapons ranged from heavy machine guns (12.7- and 14.5-millimeter) to 100-millimeter weapons. In defense of their SAM sites, the NVN deployed 57-millimeter anti-aircraft guns. This was an automatic weapon with a rate of fire of 80 rounds per minute. Under radar guidance, it could engage targets up to an altitude of 20,000 feet. Without radar, its engagement ceiling was limited to about 13,000 feet.[8] The capabilities of the 57-millimeter gun, as well as the 37-millimeter gun, complemented the SA-2 well, because any aircraft that flew below the SA-2's engagement envelope entered the range of the 57- and 37-millimeter guns. It also possessed heavier weapons, of 85- and 100-millimeter guns, but these were fewer in number and less deadly to low-flying aircraft. From March through December of 1965, 57-millimeter, 37-millimeter, and lighter weapons accounted for 90 percent of the total losses of U.S. aircraft.[9]

The guns were most often guided by another Soviet radar, which NATO had code-named "Fire Can." This system was copied from a U.S. model used in World War II. The heavier caliber 85-millimeter and 100-millimeter guns were most effective when radar guided, as they were not automatic weapons. The 85-millimeter gun had a rate of fire of 15–20 rounds per minute, and could effectively hit targets at 27,000 feet. The 100-millimeter weapon's rate of fire was 15 rounds per minute, reaching targets at 39,000 feet. Fire Can's weakness lay in its lack of a moving target indicator (MTI), which could distinguish between radar blips that were moving from those that were static. Without an MTI, the radar operator had to do this manually; this was an especially challenging task when engaging low-flying aircraft whose radar returns could easily be confused with those produced by the surrounding terrain. When faced with targets flying too low for radar guidance, the NVN would use barrage fire of 57-millimeter and 37-millimeter guns, heavy machine guns, and even automatic rifles. Rather than aiming at any particular target, this method relied upon erecting a "wall of fire" through which an aircraft would have to fly. USAF documentation stated that, "According to North Vietnamese Army manuals, a trained rifle platoon could fire a barrage of 1,000 rounds in 5 seconds or less."[10] Even very light weapons could thus prove to be effective if used properly.

The first encounters with the SA-2 ended badly for the USAF. In July 1965, the four F-4C Phantoms of "Leopard" flight were escorting a Rolling Thunder bombing mission over North Vietnam. Flying at 23,000 feet, they received a warning from a nearby reconnaissance aircraft of emissions that signaled an SA-2 battery in the area. The warning came too late, however. Before the flight could evade the NVN's weapons, two missiles closed on the F-4s rapidly. One missile detonated below Leopard 02, piloted by USAF Captains Richard Keirn and Roscoe Fobair, downing the aircraft, killing Forbair, and making Keirn a POW for the rest of the war. The other Leopard craft evaded the other missile and returned to base, where it was discovered that all three surviving aircraft had sustained damage from the missile that had destroyed Leopard 02. It was the second time that the USAF had detected an SA-2 site electronically in two days, but the first time that one drew blood.[11]

One missile had downed an aircraft, causing the loss of both crewmen, and damaged three others. The Johnson administration, wanting to deter the NVN's further use of SA-2s, gave the go-ahead for a retaliation mission against known sites. Flying low to avoid the SA-2s, the F-105 "Thunderchief" fighter-bombers encountered heavy AAA at all of their target areas. The NVN downed a total of four F-105s and damaged a fifth. As a wingman of the damaged aircraft flew in tight formation with it to assess the damage (a standard practice then and now), the two aircraft collided, resulting in two more aircraft losses. One of the SAM sites was subsequently revealed to be a decoy; no missiles or other SA-2 systems were present, but numerous AAA guns were.[12] The low-altitude approach of the F-105s made them especially vulnerable to the guns positioned at those sites. Prior to the Leopard flight incident and the disastrous retaliation mission, the USAF had done little to counter the NVN threat. The loss of a total of seven aircraft in such a short amount of time spurred on their adaptation effort in earnest.

The U.S. Air Force at the Outset of Hostilities

The main organizational division within the USAF at the outset of the Vietnam War was between Strategic Air Command and Tactical Air Command. The former, by far the dominant branch, was focused on execution of the Single Integrated Operational Plan (SIOP) in the event of war with the Soviet Union. At the time, SIOP was a blueprint for employing the U.S. nuclear arsenal in an all-out strategic bombing campaign against the Soviet homeland and Eastern Europe. Defense of the North American landmass against Soviet bombers was also a priority. In the years between the Korean War and the Vietnam War, U.S. defense policy had shifted from conventional forces to nuclear, in support of the Eisenhower administration's Massive Retaliation policy. In budgetary terms, this meant that the armed services and their subunits could best justify their existence, and stand a better chance at expansion, by focusing on the delivery of nuclear weapons. The nature of its mission meant that the USAF, and specifically SAC, had a leg up in the competition for funding. Not only did the other services receive less funding, but TAC also became a decidedly inferior partner to SAC within the USAF.

Furthermore, "bomber generals" dominated the leadership of the USAF. Epitomized by General Curtis LeMay, the architect of the strategic bombing campaign against Japan in 1945, these generals had flown combat missions in strategic bombers during World War II. They were combat veterans who were true believers in the efficacy of strategic bombing. Their worldview conformed nicely to the mindset of Massive Retaliation, and their experience facilitated their rise within the ranks. In the years before the Vietnam War, they had risen to the top of SAC, but also of the USAF and TAC. Fighter pilots and others who had significant combat experience in World War II outside of the strategic bombing campaigns were passed over for promotion or otherwise minimized in importance within the USAF.[13]

This relationship had significant implications for how the USAF was structured as it entered the conflict in Vietnam. First, TAC was underfunded. The USAF did not adequately fund weapons programs designed to fight in non-nuclear conflicts. TAC relied on avoidance of Soviet air defenses rather than degrading them through direct action. This meant that TAC aircraft went into combat without adequate countermeasures against SAMs. Second, TAC's need to justify its existence and its strategic bombing–focused leadership turned it into a "mini-SAC," and its aircraft and tactics reflected this. As a subunit, it expected to prosecute a war in central Europe or assist in SIOP by delivering nuclear weapons against Soviet targets. Because the conflict that TAC expected to fight would be a one-time, all-out Armageddon, it placed little emphasis on degrading enemy air defenses. Furthermore, bombing accuracy was not as important with nuclear weapons, so training in delivery of conventional weapons suffered accordingly.[14]

The aircraft available to the USAF at the time of Operation Rolling Thunder reflected the "mini-SAC" orientation of TAC. The F-105 "Thunderchief," in particular, is most relevant here. It was the primary ground-attack aircraft used over North Vietnam. Ironically, SAC's strategic bombers—B-52s—saw extensive service in Southeast Asia, but were deployed mostly over South Vietnam and the Ho Chi Minh Trail on "tactical" missions. B-52s did, however, see action over North Vietnam in 1972 during the Linebacker operations, which I discuss below.

The F-105 was built primarily to carry tactical nuclear weapons at low altitude in the European environment. It had an internal bomb bay (a first for a single-seat fighter), a large and powerful engine, and short wings. This combination allowed it to reach speeds in excess of the speed of sound at treetop level. The philosophy behind this design was that it could fly below the advanced SAMs deployed by the Warsaw Pact and outrun any MiGs that it might encounter on its way to and from the target. The F-105 bore the brunt of the mission load in the air campaigns against North Vietnam from 1964 to 1970. Its pilots suffered the highest casualty rates by type of fixed-wing aircraft, in part because more than 50 percent of their missions were flown against the most heavily defended targets. By the end of the their service in Southeast Asia, 60 percent of the F-105s available had been destroyed.[15]

Despite the dangerous mission and the heavy losses, the F-105 enjoyed the fierce loyalty of its pilots. The speed of the aircraft made it difficult for MiGs to engage. While the F-105 suffered heavily from NVN air defenses, the fact that it flew so fast prevented the losses from becoming even more severe, because its high speed meant that it could outrun its enemies.[16] In addition, its heavy construction meant that it could sustain more damage and keep flying, at least long enough for its pilot to eject more safely. In one instance, a SAM blew off several feet of a two-seat F-105's wing, but the craft stayed in the air long enough to carry its crew over the sea where they could eject and be rescued.[17]

While the F-105 was generally quite rugged, it had its vulnerabilities. Because it was considered most likely to face the large and sophisticated SAMs, little emphasis was placed on designing redundant systems, as such a hit was not likely to be survivable. The most significant result of this was that the aircraft's backup hydraulic system was positioned right next to the primary system; thus, one lucky hit could disable both systems at once. Without hydraulics, the aircraft was uncontrollable. As demonstrated by the 60-percent loss rate, rugged construction only carried the F-105 so far. The optimal mission profile for the aircraft—low altitude—played to the advantage of NVN air defenses, especially its AAA.[18] USAF war planners had factored out this vulnerability to AAA by reasoning that "the Soviet border was simply too large to surround with a practical air defense system. The tacticians believed that the F-105 could exploit the large gaps in AAA placement on its way into the Soviet Union."[19] Of course, this reasoning did not work when applied to North Vietnam, because that country's border was much shorter, and it tended to place its air defenses near potential targets rather than attempting to seal off the country from air attack.

Operation Rolling Thunder

The goals of the Rolling Thunder campaign were to coerce the North Vietnamese to stop infiltrating its forces and supplies into South Vietnam, and to force a negotiated settlement to end the war.[20] According to Pape, the United

States pursued three different strategies in trying to achieve these goals. First, civilians in the Johnson administration advocated an escalation strategy, threatening North Vietnam's industrial base but not inflicting so much damage that it would be destroyed. This would simultaneously hurt something the North valued, but keep that object healthy enough not to negate the threat—"keeping the hostage healthy," because a dead hostage is no longer valuable. Furthermore, the campaign would be designed to inflict escalating damage on the North Vietnamese that would demonstrate U.S. resolve to increase the punishment it was willing to impose. The final component of this strategy was to explicitly couple military force with diplomacy to facilitate a settlement. This required careful direction of military actions by civilians in order to ensure that the correct signals were being sent.[21]

The second proposed strategy, which was favored by the USAF, was a more intense one designed by USAF Chief of Staff General Curtis E. LeMay and his successor, General John P. McConnell. It called for air strikes that would destroy the North's industrial base rather than hold it hostage. By hitting some 94 targets in quick succession, this campaign would "create panic and divert Hanoi's attention away from the South toward its own internal problems," undermining civilian morale.[22]

The third proposed strategy was to isolate the South Vietnamese battlefield by interdicting supplies and manpower flowing from north to south. By focusing on the military vulnerability of this long supply line, the advocates of this strategy—Joint Chiefs of Staff Chairman General Earl G. Wheeler and Army Chief of Staff General Harold K. Johnson, among others—hoped to deprive the insurgency in the South of supplies it needed to continue its operations.[23]

Rolling Thunder employed all three of these strategies in turn. In the spring and summer of 1965, the United States tried the gradual escalation and hostage-taking strategy. Air interdiction, or trying to cut off supplies to the Southern insurgents, was the strategy from the summer of 1965 to the winter of 1966–1967. From spring to fall of 1967, the Johnson administration reacted to the failure of the other strategies by deploying the more intense campaign favored by LeMay. This phase was so intense that there was little room for escalation, as few targets were left. Other than bombing and mining North Vietnam's ports, a land invasion was the only option for further escalation. After the Tet Offensive in early 1968, however, the war lost U.S. public support and deescalated from April to November 1968, when the Johnson administration prohibited attacks on targets at an ever-enlarging distance from Hanoi. This period saw a return to interdiction as Rolling Thunder wound down.[24]

The targets that the USAF typically attacked were in keeping with the strategies of coercion employed by the United States. They included the few industrial assets that North Vietnam possessed, such as power generation facilities and small factories. Later, bridges, supply dumps, and railroads became the focus of interdiction efforts. Poor weather over North Vietnam frequently diverted strikes originally sent after fixed targets toward the interdiction mission: when the notoriously bad weather obscured visibility over the primary target, the F-105s would patrol for targets of opportunity along roads. The F-105s flew from central Thailand, and received support from F-4 fighter wings based near the eastern Thai border and at Da Nang in South Vietnam, as well as search-and-rescue units in the same areas.[25]

While Pape asserts that an analysis of the Pentagon Papers indicates that few targets were then off-limits to U.S. forces, USAF documents indicate otherwise.[26] The "ring" around Hanoi and Haiphong harbor within which U.S. aircraft were prohibited from engaging targets certainly shrunk, especially around the harbor. The port itself, however, through which most of North Vietnam's military hardware passed, was "never mined lest a Soviet ship be sunk."[27]

The U.S. Air Force's Adaptations

As NVN air defenses quickly became a truly integrated system, the USAF pursued a gradual adaptation. James Brungess goes so far as to say that the reaction was piecemeal in nature and fundamentally unintegrated, thereby putting the USAF at an inherent disadvantage. According to his argument, the USAF's approach to the air defense suppression was reactive and based upon neutralizing air defense targets in isolation, rather than treating them as part of a greater system. A better approach to the problem, he suggests, would have been an attack on the system as a whole and within a short period of time, designed to engage the entire NVN air defense network, rather than its various nodes defending specific targets. The USAF was very good at degrading the SAM, but AAA and MiG interceptors still took their toll throughout the war because the NVN radar network as a whole operated unmolested. Such an approach, however, was incompatible with what the political authorities permitted the USAF to do over North Vietnam. Decision-makers and strategists allowed the USAF to attack NVN air defenses, but only in a piecemeal or individually targeted manner. Both political considerations (Johnson administration worries over war escalation) and doctrine problems (the USAF viewed the problem in terms of targets, rather than networks) prevented a total-system approach to IAD degradation. The problem was one of tempo; an ideal approach would have been to attack the system as a whole quickly, precluding its ability to recover, but political considerations dictated a slower approach.[28]

The first USAF adaptation was the "hunter-killer" team concept. Aircraft such as the RF-101, a reconnaissance version of the F-101, and F-4C would fly over areas suspected to have SA-2 sites. Once they goaded a SAM site to engage them, these aircraft would note the location of the site and communicate it to alert aircraft. Photo reconnaissance was required to confirm the presence of a SAM site at a given location. The alert aircraft, F-105s waiting at their airfields in Thailand with conventional munitions, would launch and attack the site. The system was feasible in theory, but the lag time involved in the communication between its "hunter" and "killer" elements made it completely ineffective: "To process the film, interpret the pictures, and dispatch a strike took from 6 to 8 hours."[29] This gave the NVN ample time to pack up the entirety of an SA-2 battalion and move it to another site. Because the USAF's alert system was ineffective and wasteful, it was quickly discontinued.[30]

The USAF also attempted to use drones to ferret out SAM sites. The drones would fly through NVN airspace, tricking enemy radars into transmitting. Electronic warfare aircraft, such as the EB-66, would determine the position of the site using the radar signals, and transmit the location to the alert aircraft. This method decreased the reaction time to two to three hours, but that still turned out to be more than enough time for the NVN to move the SAMs before the strike arrived.[31]

Losses to NVN SAMs prompted the USAF to form a special commission tasked to address the problem. In early August 1965, USAF Chief of Staff General John P. McConnell directed that the commission discuss and implement workable adaptations to the SAM. Brigadier General Kenneth C. Dempster, Director of Operational Requirements and Development, headed the new commission, thereafter called the Dempster Commission. It started the next phase of the adaptation process by recommending the acceleration of development of jamming pods under the Quick Reaction Capability (QRC) system. This system avoided the need for excessive testing of new equipment that combat needs required and aimed to get that equipment to fighter wings more quickly. Jamming pods, designed to be carried by fighter aircraft, had been under development since 1961, but until the appearance of the SAM in Vietnam, they had been underfunded by TAC. The pods were designed to broadcast electronic signals that interfered with the radars associated with the SA-2 and radar-guided AAA. More detail concerning the pods will follow below, but at this point it is sufficient to state that the relatively slow-moving program to develop them was given a significant boost at this meeting.[32]

Furthermore, the Dempster Commission set requirements for special warning gear designed to alert pilots to nearby SAMs and their associated sensors. Dempster wanted such a device in every combat aircraft that flew over North Vietnam. He directed a competitive demonstration of available devices, which was held at Eglin Air Force Base in Florida. Applied Technology's Vector sensor won the three-day competition, and the USAF selected it for production.[33] Designated APR-25/26, the equipment detected the SA-2's tracking and guidance radars and displayed the direction from which the threat originated, allowing the pilot to take evasive action.[34] This information was displayed upon a small cathode ray tube (CRT) in the form of spikes radiating out from the center of the screen. A solid, long line represented the most urgent threat to the aircraft, while those that were shorter and dashed signified that a SAM had not yet been launched. This was the first warning gear carried into combat by USAF fighters. The various systems used for this purpose were classified as Radar Homing and Warning System (RHAWS) gear.[35] In a further testament to Dempster's influence, Applied Technology placed orders for parts to construct one thousand more APR-25/26s based solely on the general's word that a large order was coming soon, before a contract actually existed. The USAF's contract arrived at Applied Technology just in time to cover its expenses.[36]

The other most significant product of the Dempster Commission was the development of the "Wild Weasel" mission.[37] The term "Iron Hand" is a more generic one representing missions aimed at suppressing enemy air defenses, to which any USAF aircraft can be assigned. Wild Weasel was a subset of Iron Hand specifically aimed at SAMs and radar systems and required a specially equipped aircraft. As David R. Kendziora writes, "'Wild' accurately described the atmosphere of the combat and the spirit of the crew members who volunteered to preempt the new ground fire. 'Weasel' referred to the bloodthirsty mammal that cunningly pursues vermin."[38]

The hunter-killer concept's main failing was the extremely slow reaction time once a SAM site was detected. Combining the hunter and killer roles into one aircraft would negate this disadvantage. The Wild Weasel aircraft would fly into a target area ahead of the main strike, seeking out enemy surveillance and guidance radars, and marking them for accompanying Iron Hand aircraft, which would destroy the sites with bombs and rockets. The Wild Weasel would use special electronic equipment to detect and interpret electronic transmissions. Not only did this mission require a daring pilot willing to expose his aircraft to enemy fire, it also required a highly trained Electronic Warfare Officer (EWO) to decipher the radar signals and vector the pilot to the target site. Fortunately, the USAF had plenty of well-trained personnel for the EWO role available, because SAC bombers and reconnaissance aircraft all had these officers as regular components of their crews; however, flying into the teeth of enemy air defenses and willingly exposing oneself to AAA and SAMs seemed counterintuitive at best. When confronted with these challenging mission parameters, an experienced officer who was tapped to lead the program reacted with the statement, "You gotta be shittin' me." The commander did not stay with the program, but the phrase stuck, often featured prominently on Wild Weasel aircrew uniform patches as the acronym "YGBSM."[39] Wild Weasel pilots and EWOs took pride in the harrowing nature of their mission; from its inception the program enjoyed a high *esprit de corps*.

The Wild Weasel program integrated an aircraft, special electronic equipment, and later special weapons designed to home in on enemy radar emissions. While certainly not representing the entirety of the USAF adaptation, Wild Weasel embodied each element of it. There have been multiple types of aircraft that have served as Wild Weasels, each more advanced than the last. A short description of the increasingly advanced systems employed by the USAF follows.

The first Wild Weasel aircraft, designated Wild Weasel I, was based upon the two-seated F version of the F-100 Super Sabre. The aircraft had a range of approximately 1,500 miles, with a maximum airspeed of Mach 1.3 at 40,000 feet. It carried a mix of 1,000-pound bombs and 2.75-inch unguided rockets, in addition to two 20-millimeter cannons.[40] Its most important piece of equipment was the IR-133, which basically fulfilled the same function as the APR-25/26 RHAWS. The IR-133, however, received information from better sensors, and displayed it on a larger six-inch CRT scope. It also had a computer that could identify and classify the radar signals it detected. A CIA-led operation, United Effort, enabled the United States to identify and classify SA-2 electronic signals. In February 1966, a modified drone succeeded in collecting and transmitting to a waiting RB-47H "the full sequences of guidance and downlink signals."[41] This data was directly applied to development of the Wild Weasel I RHAWS gear. It provided enough information to the EWO that the Wild Weasel could hunt the SAM site, rather than just avoid it as with the APR-25/26.[42] The first four of the newly modified F-100Fs passed flight tests in September 1965, flew to Eglin Air Force Base for weapons tests in October, deployed to Korat Air Base Thailand in November, and entered combat in December.[43]

On a typical Iron Hand mission, a Wild Weasel F-100F would fly accompanied by three or four F-105s. The F-100 EWO would interpret the signals received by the IR-133. No range-to-target information was given by this system (or any other until the Wild Weasel IV, which entered service after the Vietnam War was over), so even with the special gear, hunting SAMs was a tricky proposition.[44] Once the Wild Weasel found a site, it would use its 2.75-inch rockets to visually mark the site for the Iron Hand F-105s. The F-105s would then use their own bombs or rockets to destroy it. The F-100F/F-105 team achieved its first SAM site "kill" on December 22, 1965, using rockets and cannon fire.[45]

The Wild Weasel I suffered from several disadvantages, as one might expect from a stop-gap measure. The fact that it was a different airframe than the aircraft it was supporting was the source of most of the problems. The F-100F's range was significantly less than that of the F-105. The F-105 was also much faster—capable of Mach 2.1 when flying at 40,000 feet—and this caused coordination problems in mission planning. Not only would the Iron Hand F-105s have to fly slower than the Wild Weasel aircraft, but also the strike packages it supported could not be escorted throughout their entire mission if they did not want to sacrifice speed. The Wild Weasel would have to get airborne first, and proceed to the target ahead of the strike aircraft. En route to the target, the F-105 Iron Hand aircraft would have to weave back and forth across the F-100F's flight path in order to compensate for the slower cruising speed of the latter.[46] In terms of aircraft survivability, the F-100F's poor maneuverability and limited speed meant that it was difficult for it to evade anti-aircraft threats.[47]

Almost as soon as the F-100F was deployed on Wild Weasel missions, the USAF searched for a replacement. Wild Weasel II was an F-105F with wing-mounted electronics. Its equipment configuration was such that the wings needed reinforcement and its airspeed had to be kept below 300 knots to limit airframe vibration. These restrictions made the aircraft unsuitable for combat, and Wild Weasel II was thus quickly surpassed by the Wild Weasel III, which had internally mounted electronics. In January 1966, Dempster secured funding for the Wild Weasel III, which was based on the F-105F platform.[48] The F-105F was suitably modified and deployed to Thailand in spring of 1966. Like the F-100F, it was a two-seat aircraft. The warning gear was improved to present data to the EWO in a clearer fashion. In addition to the SAM radars, it could now detect ground-controlled intercept (GCI) radars, which North Vietnam began to use to provide targeting data to the SAM sites. In addition to visual cues, the EWO received audible signals in the form of variously timed beeps that enabled him to interpret the threat. The improved IR-133 was redesignated ER-142.[49]

The arrival of these F-105F's marked the end of those Iron Hand formations designed to compensate for the difference in speed between the Thunderchief and the F-100F. Hunter-killer flights came to consist of two two-plane elements, at least one of which was led by a Wild Weasel. They generally approached the target in the same loose fingertip formation as a flight of strike aircraft, then separated to cover a broader area.[50]

With the upgrade to the F-105F, the Wild Weasel also received its first specialized weapon: the anti-radiation missile (ARM).[51] During the Vietnam War, the most common of these was the AGM-45 Shrike. The Shrike was the first U.S. ARM. Specifically designed to counter the SA-2, this weapon started out as a U.S. Navy (USN) program. It was built using an AIM-7 Sparrow air-to-air missile frame, but used a larger warhead and smaller rocket motor than that missile. Its seeker head homed passively on the signals generated by the SA-2's Fan Song radars. Its range was 17–19 miles. In order for the Shrike to work, the pilot pointed the aircraft at the target, allowing the missile's seeker head to acquire the radar signal. Once the missile was launched, the enemy radar had to continue to broadcast until the missile hit it. Otherwise, it would assume a ballistic trajectory, as it did not have any memory circuit that could "remember" where the target was.[52] Employing Shrike was much more complicated than generally known:

> The weapon's sensor head reacted to signals over a specified frequency band and could home on any radar that functioned within this band. A proximity fuze [sic] detonated the 63-kilogram (140-pound) warhead. The pilot launching a Shrike had to fly directly toward the radiation source and loft the missile so that its ballistic track would bring it close enough to the target for the homing system to take over. Some 10 seconds after launch, the rocket motor burned out and guidance ceased, but if released precisely the weapon could strike within 20 feet of the radar antenna against which it was directed.[53]

Furthermore, establishing the ballistic trajectory required a calculation that was impossible to accomplish in the short time available:

> Because the electronic gear on board the Wild Weasel could not determine the precise range to target, the crew had to use some other method of getting the Shrike within homing distance of the enemy

> radar. A typical launch maneuver consisted of diving toward the transmitter until the missile's radiation seeker had locked onto the proper azimuth and elevation, then pulling up and lofting the Shrike toward its target. A reference table told the crew what loft angle to use for a given speed, dive angle, and altitude. This aiming system had two built-in weaknesses. First, the Wild Weasel was hurtling through the air as the crew made these calculations and was closing with the radar transmitter. Second, the table being consulted was based, for ease of computation, on the assumption that the target lay at sea level; therefore, if the radar was transmitting from a hilltop or ridge line, the loft angle would be incorrect.[54]

Out of 107 missiles fired from mid-April 1966, only one was confirmed to have actually destroyed a radar transmitter, while 38 were considered "probable" hits. The missile's small warhead was in part to blame, because it necessitated a direct hit to inflict substantial damage. The complicated method of delivery was more significant, however. Also, North Vietnamese radar operators quickly learned to recognize when they were about to be the target of a Shrike, and addressed their vulnerabilities to ARMs:

> The enemy found that an acceptable way of countering Shrike was to reduce transmission time by relying on acquisition radars or ground observers to supply the course and speed [of] approaching American formations. Instead of transmitting for 10 or 12 minutes, Fire Can and Fan Song could remain on the air for 3 minutes or less. The North Vietnamese also learned to recognize hunter-killer teams—usually four planes—in contrast to the larger strike formations, and to avoid using gun-laying or missile control sets while Iron Hand was nearby. Radar operators shut down as soon as they detected a Shrike launch, which could be seen on the scope or reported by observers posted in the vicinity, thus depriving the homing device of a target.[55]

Thus, the fact that the Shrike was specifically designed to counter the Fan Song system was one of its greatest weaknesses: as the NVN found ways to limit that system's transmissions, the usefulness of the Shrike ARM decreased. These counter-adaptations also limited what the Wild Weasel crews could do with their own detection gear. Without sufficient time, they could not make the necessary calculations and line up a shot in time. They soon started to rely on a "best guess" of the range to target, and launched a Shrike in the hope of forcing the radar off the air, causing the SAM to go ballistic and miss its target.[56] The USAF responded with the following strategy:

> Despite [these] limitations . . . crews did make effective use of the antiradiation missile. Since the enemy radar operators would reduce their transmissions in order to minimize the danger from Shrike, Iron Hand flights used the threat of this missile to deter the North Vietnamese from tracking accurately the strike formations. To make this menace more credible, the Wild Weasels would feint toward suspected radars, or sometimes actually bomb them, and promptly attack any Fan Song or Fire Can bold enough to begin transmitting. Occasionally, the Iron Hand teams disguised themselves by flying close to the strike force, hoping to be mistaken for part of it, in order to lull the enemy into going on the air long enough to provide a target for Shrike.[57]

Even with the improved aircraft, the mission was a dangerous one. By mid-August 1966, of the original five F-105F aircraft deployed to Korat, Thailand, only one was still operational. None of the original six at Takhli Royal Thai Air Force Base (RTAFB) were operational. After 48 days of operations at Takhli, the eight crews assigned had suffered four killed in action, two wounded, two prisoners of war, and one transfer. Only four officers eventually finished their 100 missions. Two went home at 80 missions, which was legal but early, and an option rarely taken by aircrews in general.[58]

Part of the danger originated with the tactics employed by the USAF. Standard practice until mid-1968 was for the Iron Hand flight (one or two Wild Weasels accompanied by two or three Iron Hand aircraft) to precede the strike force into the target area by as much as five to seven minutes. The small size of this flight made it clear to NVN radar operators that it was on a SEAD mission, and that they should ignore it in favor of the follow-on strike force. In addition, the vanguard Iron Hand flight was vulnerable to interceptors that might be in the area. In mid-1968, these tactics were revised, and the Iron Hand flight preceded the strike by only one minute. An additional Iron Hand flight accompanied the fighter-bombers, on the flank of the formation from which the most likely threat of radar controlled anti-aircraft fire might come.[59]

The USAF had recognized the shortcomings of Shrike, and fielded a better missile in March 1968 in the form of the AGM-78 Standard ARM (STARM). It never replaced the Shrike, but augmented it on Wild Weasel aircraft. Like Shrike, the missile was developed by the USN, which used it for shipboard air defense. It was essentially a RIM-66 Standard SAM modified with a different seeker head and attachments for aircraft to carry it. The STARM represented a marked improvement over Shrike. Launched from 40,000 feet, it could achieve a theoretical range of 60 to 70 miles, although most Wild Weasels flew at about 24,000 feet, so operationally its range was shorter. It was still much longer-ranged than Shrike, however. Its warhead was a larger 99 kilograms (219 pounds), which gave the STARM more destructive power. Most importantly, the missile did not have to be pointed directly at the target radar in order to home in on its signal. Furthermore, it contained a memory circuit that gave it the ability to head toward the target even after it stopped transmitting.[60]

Shrike remained in service for two important reasons, however. First, the STARM was twice the weight of Shrike, and excess weight affected aircraft performance. Second, while Shrike cost about $2,000 per missile, STARM cost about $200,000 each. The latter cost was so prohibitive that aircrews were not allowed to fire practice shots in training, as they did with the Shrike.[61] When the Department of Defense requested evidence of success of the USAF's electronic warfare, no hard data was available. Therefore, the USAF directed its Air Force Special Communications Center (AFSCC) to systematically evaluate its electronic warfare efforts in Southeast Asia.[62] This evaluation concluded that the presence of Wild Weasel aircraft "reduced SAM firing rates considerably and sometimes by as much as 90 percent."[63]

For the entire Rolling Thunder campaign, Wild Weasel aircraft had been parceled out among the Thailand-based F-105F squadrons.[64] This meant that each squadron had only one or two Wild Weasel aircraft. The result of this arrangement was that the specially trained pilots and EWOs rarely flew Wild Weasel aircraft. The USAF corrected this by consolidating the aircraft and crews into Wild Weasel squadrons in July 1970.[65] Prior to this point, the USAF had instituted the training, developed the necessary equipment, and the community had fostered its strong *esprit de corps*, but the resources necessary for the SEAD mission were being used inefficiently. Creation of a subunit specifically tasked with countering ground-based air defenses was thus completed too late for Operation Rolling Thunder (but well before Operation Linebacker).

Besides active countermeasures like the Iron Hand/Wild Weasel teams and ARMs, and strictly "passive" RHAW gear, the USAF used a variety of electronic adaptations to counter the NVN's mode of operation. Because the SA-2 and guided AAA fire relied upon radar, electronic jamming became an important element of the USAF response. This came in the form of special jamming pods deployed on strike aircraft, and specialized jamming aircraft like the EB-66.

The EB-66 was a reconnaissance and electronic warfare aircraft based upon the airframe of an obsolescent light bomber, the B-66, which was powered by two turbofan engines and crewed by six personnel. It was first used by the USAF in May 1965 to gather intelligence about NVN air defenses, specifically looking for Fan Song activity. It was an EB-66 that first detected a Fan Song signal a day before, then later seconds before Leopard 02 became the first SA-2 casualty.[66] There were not enough EB-66s in Southeast Asia from April to October 1965 to fully support Rolling Thunder's jamming operations, so the USAF previously relied upon other aircraft. During early Rolling Thunder operations in spring and summer of 1965, it was Marine Corps aircraft that accomplished the jamming, using the EF-10 Skynight. They were used to interfere with Fire Can, early warning, and GCI radars. Only multiple aircraft working together were effective in jamming Fire Can and Fan Song. These aircraft, however, were of 1940s vintage, had limited range without the capability of refueling from USAF tankers, and suffered from poor performance in the air. As soon as the NVN started to use SA-2s in numbers, these aircraft had outlived their usefulness: "By late 1965, the proliferation of missile sites and the increased skill of SAM crews had forced the Skyknights to orbit too far from most inland targets for effective radar jamming."[67]

By October 1965, the USAF had enough EB-66s to support Rolling Thunder, and eventually to equip two electronic warfare squadrons at Takhli RTAFB by 1968. The EB-66 had much greater jamming capability than that of the Skyknights, as well as being more survivable. Two versions, the EB-66B and the EB-66C, were used. The C model specialized in intelligence-gathering (although it still had more jamming transmitters than the Skyknight), while the B model specialized in jamming:

> The EB-66Bs made their combat debut in October 1965 and soon demonstrated that the very jamming power that enabled them to close with the target could also be a defensive liability, for the electronic noise emanating from the plane interfered with its radar warning equipment. As a result, the EB-66B and C had to work together. The broader and stronger jamming barrage laid down by the Bs afforded better protection against the SAMs, allowing the C models to remain well beyond the 17-nautical-mile range of the Guideline missile. In practice, therefore, while one or two B models

> orbited within 15 nautical miles of the target, an EB-66C remained safely beyond missile range, providing SAM warning and ensuring that the jamming barrage blanketed those frequencies the enemy was using.[68]

The EB-66s would orbit in racetrack patterns while the strike forces would fly in and out of the target area. The USAF typically flew these patterns in the Gulf of Tonkin off Haiphong harbor, over the western portion of North Vietnam, and also northwest of Hanoi behind a mountainous area dubbed "Thud Ridge." The latter was so named because it formed a convenient approach route (running from the northwest to southeast, pointing in the general direction of Hanoi) for strike aircraft, predominantly F-105F Thunderchiefs, nicknamed "Thud" by their pilots. While the F-105Fs used the ridge to mask their approach from Hanoi's air defenses, the EB-66s would orbit at its northwest tip. There were also standard orbiting points around North Vietnam's panhandle, which typically were over the Gulf of Tonkin, on the southern side of the demilitarized zone (DMZ). The EB-66 operators had to take into account the pattern of the jamming signals that the transmitters emitted, as it was not a uniform circle in shape. The signal was much weaker to the front and back of the aircraft, so that the aircraft had to maintain a broadside attitude toward the target. In order to maintain the integrity of the jamming barrage, this meant that more than one aircraft had to work together.[69]

As the NVN expanded its air defense network, even the EB-66s became vulnerable. The SAM network expanded toward the EB-66 orbiting areas in the south and west of North Vietnam, and the jamming aircraft had to retreat farther south for safety. They soon retreated so far, however, that the jamming signals became ineffective where they were most needed: over the targets: "The EB-66 unfortunately was incapable of accompanying strike forces into heavily defended areas and therefore could not effectively assume countermeasures responsibilities. In the words of Col. Arthur D. Thomas, what the Air Force needed was 'a penetration aid that would defeat SAM and radar systems throughout the mission.'" That meant kicking the development of self-protection jamming pods into high gear.[70] USAF documentation states that "fortunately, the introduction late in 1966 of jamming pods for individual fighter-bombers and reonnaissance planes offset the effect of withdrawing the EB-66's from Thud Ridge."[71]

According to one general, the USAF entered the Vietnam War "almost electronically naked, and remained that way for the first year of combat." It was aware of the threat of the SA-2, and knew that its major potential adversary, the Soviet Union, had an extensive air defense network. Its capability to operate in an environment filled with SAMs and radars, however, was limited to SAC. TAC was slowly developing countermeasures for its aircraft, but the process was incomplete as Operation Rolling Thunder began. The USAF's preoccupation with SIOP meant that the ability to counter air defense radars did not filter down to TAC until NVN air defenses drew first blood. TAC's funding simply did not allow for the organization to expend the necessary resources for what, at the time, were expensive countermeasures whose capabilities were uncertain.[72]

> Prior to the Vietnam war [sic], TAC was confident that it was developing and gradually acquiring countermeasures gear that would disrupt modern radar-controlled defenses. The watchword at the time was "quality rather than quantity." Once the Southeast Asia fighting began, however, North Vietnam's defenses improved with stunning speed, and TAC found itself struggling to keep pace.[73]

The Dempster Commission, which originated the Wild Weasel program, also drastically sped up the development of jamming pods under the QRC system. Combat conditions clearly warranted that jamming pods be more quickly developed and sent to Southeast Asia, and the USAF set about doing so.[74] In development since 1963, the self protection pod QRC-160 measured about ten inches wide, a little over eight feet long, and weighed 200 pounds. Mounted on a pylon that would otherwise carry ordnance or an extra fuel tank, the pod was powered by a propeller-driven generator and would broadcast signals specifically tailored to confuse the Fan Song system.[75]

By January 1966, the QRC-160 pod was ready for testing under combat conditions. The USAF had preliminary answers as to whether or not it was rugged enough to withstand combat operations and could degrade the effectiveness of the Fan Song and Fire Can. General William W. Momyer, the chief of the Seventh Air Force (the USAF organization responsible for prosecuting Rolling Thunder), was convinced of the device's potential, and directed that it be tested in combat.

September 1966 saw the launching of Project Vampyrus for that purpose.[76] From September 26 to October 8, 1966, F-105Fs from Takhli RTAFB flew nineteen test missions against targets in North Vietnam. The test flights consisted of four F-105Ds mounting the pods (two for each aircraft, which was the standard F-105 deployment), supported by EB-66s. The EWOs on board the EB-66s "first verified the number, location, type, and transmission characteristics of radars protecting the targets, then observed how these transmitters reacted to the pod-carrying Thunderchiefs. When necessary, the EB-66Cs jammed radars outside the frequency range covered by the self-protection pods."[77]

Sometimes pods were not available for one or two of the planes in the formation. On these missions, the different experiences of the protected planes versus the unprotected craft demonstrated how effective the countermeasures could be. Planes that had no pods or pods that suffered mechanical breakdowns soon found themselves the subject of particularly intense AAA and SAM fire, while elements of the same flight with operational pods "experienced only light to moderate fire."[78]

The nature of the pod's operation meant that the formation in which the four F-105s flew was important to their effectiveness. Their small size limited pod-jamming power. Therefore, one of the objectives of the Project Vampyrus tests was to determine the best way to arrange pod-equipped aircraft formations. The tests found that a separation of about 1,000 feet vertically and 1,500 feet horizontally, staggered above and behind the lead plane, worked best. This way, the pods generated a large area of noise that masked all of the aircraft in formation, but they were not so close together that pilots had to be inordinately concerned about colliding with their wingmen, and could thus still scan for targets and threats effectively. There was a slight variation in separation distance employed by the two F-105 fighter wings based in Thailand, but it did not have appreciable effect. With the success of the Project Vampyrus tests, the pod entered full production and was redesignated ALQ-71: "By January 1967, the Seventh Air Force had enough ALQ-71s and mounting brackets to begin equipping F-4 Phantoms as well as F-105Ds."[79] The F-4s were the principle fighter escort of the Rolling Thunder strikes, and also accomplished some air-to-ground operations. These pods gave enhanced protection to the fighter-bombers, and increased their accuracy by reducing the need for evasive action over the target.[80] Overall, the introduction of the pods greatly improved survivability. Aircraft losses in the six months after their introduction amounted to less than a third of that in the six months prior.[81]

Like almost all countermeasures, the jamming pods had their vulnerabilities. They were originally designed for high-speed, level bombing runs in what was likely a single strike during a nuclear war. The rigors of the Rolling Thunder campaign meant that the pods operated daily in tropical conditions with substandard maintenance. The QRC system bypassed normal procurement procedures and thus regular USAF maintenance personnel were not familiar with them. Specially trained personnel and spare parts were necessary. Because of these factors, pod reliability therefore suffered.[82] Within eight to ten miles of a SAM site, signals from a Fan Song system could overpower the pod's jamming, in what was called "burn through." Also, because of the aircraft's antenna propagation pattern, jamming coverage decreased markedly during maneuvers, especially in steeply banked turns when the strongest signal might not blanket every radar-controlled weapon that could track the target and open fire.[83] This meant that during combat maneuvers, holes in the jamming pattern would occur, giving windows of opportunity to SAM operators.

Because the pods transmitted at the frequency of Fan Song and Fire Can, they had the potential to interfere with Wild Weasel missions that hunted those same signals. Iron Hand aircraft in particular had to be cautious about operating the pods because of their close proximity to the Wild Weasel aircraft. The pod could reduce accuracy of Shrike and interfere with the RHAWS gear.[84] The F-4s were initially equipped with just one mounting bracket for the pods, as opposed to the F-105s' two. Coupled with the frequent turns required during combat, this meant that the F-4 had half the jamming power of the F-105 and greater instances of gaps in jamming coverage, which together meant that "the noise barrage scarcely inconvenienced skilled radarmen."[85]

North Vietnam continuously looked for ways to undermine U.S. countermeasures:

> In November 1967, SAMs scored eight kills within just 4 days, despite the self-protection pod. An investigation team flew to Southeast Asia and discovered that a combination of faulty radar bombing technique and . . . North Vietnam's defenders had been responsible for the startling losses. Some Fan Song radars were transmitting on a slightly lower frequency, thus escaping the jamming barrage. Also, Air Force planes tended to bunch up in order to obtain a more compact bomb pattern during missions controlled by ground-based radar, and in doing so they simplified the task of tracking and aiming at the jamming source.[86]

In other words, the pods only worked well if the aircraft were separated enough so that they appeared to be a rather large, undefined blip on enemy radar screens. If they were too close together, they presented a more defined target that could be engaged. The NVN also used passive means of tracking

U.S. aircraft by relying on the jamming source rather than a radar return. The jamming would give away the general position of the strike force, then two widely separated radar sets would transmit briefly, using triangulation to verify the range. Thus, "a really expert Fan Song operator could avoid using the tracking beam and guidance signal until the SAM had risen from the launcher. Although there was some sacrifice of accuracy, these tactics reduced the impact of the ALQ-71, decreased SAM warning time, and minimized exposure to antiradiation missiles."[87]

Operation Linebacker

After the Johnson administration suspended the bombing of North Vietnam in 1968, U.S. aircraft did not return in force over the North until 1972. When they did so, the intensity of Operation Linebacker was above that yet seen during the entirety of the Vietnam War. Many, especially in air-power advocacy circles, attribute the success of these operations to the fact that the "gloves came off." Linebacker had few restrictions on its targets or operational tempo, and instead of the relatively unlimited goals of Operation Rolling Thunder, the 1972 campaigns had very focused goals. A brief description of Operation Linebacker follows for contextual purposes.

By 1972, the Nixon administration's policy of "Vietnamization" was in full swing. This policy was to turn over the burden of counterinsurgency operations to the South Vietnamese government and armed forces, while steadily reducing the U.S. presence there. Remaining U.S. forces would provide technical assistance and advising, logistical support, and air support. This meant that the United States then had a greatly reduced ground force in South Vietnam. North Vietnam thus saw an opportunity for a quick victory in 1972, and switched from an insurgency strategy to one of conventional land invasion. When it became clear that South Vietnamese forces could not stop the North's invasion on their own, the United States stepped in with air power, both in support of the battle in the South and against the attacker's ability to sustain the offensive by striking targets in the North.[88]

At first the United States responded to the North's offensive along the lines of Operation Rolling Thunder, using limited demonstration-type attacks in a precursor operation called "Freedom Train." Nixon increased the level of U.S. involvement quickly, however, quadrupling the number of B-52s and quintupling the number of tactical aircraft in the theater. The U.S. strikes crept steadily from just north of the DMZ toward Hanoi and Haiphong. Whereas the bombing of the Ho Chi Minh Trail and the interdiction mission of Rolling

Thunder had proven ineffective, lines of communication in the North were much more vital to a conventional ground offensive, and thus they were vulnerable to air power. Linebacker I was primarily an interdiction campaign, and it worked. The North Vietnamese Army (NVA) soon had to halt its offensive, because U.S. aircraft were pummeling its units in the South directly, and weakening them indirectly with strikes on their supply lines.[89]

Linebacker I forced North Vietnam to the negotiating table with the United States when the North realized that a conventional ground offensive would not produce a victory. South Vietnamese intransigence stalled the subsequent negotiations, however, which caused the North Vietnamese to leave the talks. Linebacker II, dubbed the "Christmas Bombing," used large numbers of SAC B-52s to attack numerous targets in the North. North Vietnam had rebuilt much of its logistical infrastructure since the cessation of Linebacker I, so the target sets of the two operations were very similar. While Linebacker I was a six-month operation, however, Linebacker II was executed in less than two weeks. This operation again demonstrated the capability of U.S. air power to make North Vietnam's offensive impracticable, which led to further negotiations and an end to U.S. involvement in the war.[90]

Further Iterations of the Innovator-Adapter Relationship

The USAF used primarily F-4 fighter-bombers and B-52s in support of Linebacker I operations. The new F-111 bomber made its combat debut during this time as well. Since the end of Operation Rolling Thunder, the USAF had developed laser-guided bombs (LGB), and used them extensively in the interdiction mission, especially for bridges. The F-111 figured prominently because of its bombing accuracy by both day and night.[91] USAF and USN aircraft also mined Haiphong harbor.[92]

For Linebacker II, SAC B-52s provided the lion's share of USAF striking power, supported by Iron Hand/Wild Weasel, tanker, jamming, chaff, and interceptor aircraft. The SAC aircraft proceeded to their targets in "cells" of three. The twelve-day campaign can be broken down into three phases. From December 18 to 20, three waves of B-52s struck multiple targets simultaneously at the same time each night. From December 21 to 24, there was only one wave per night, striking fewer targets. Between December 26 and the operation's termination on December 29, the single-wave pattern continued, but with more planes taking part and striking more targets.[93]

From 1968 to 1972, the North Vietnamese took advantage of the long pause in the bombing to strengthen their air defenses.[94] In addition to increasing the numbers of SA-2s available, the NVN fielded an improved version with a moving target indicator (MTI) and optical tracking. The former distinguishes moving radar contacts from stationary ones rather than requiring the operator to do it manually, greatly simplifying the task. The latter enabled the Fan Song to engage U.S. aircraft without as many electronic emissions as radar guidance. The only signals required were to communicate steering instructions to the missile that were relayed from the optical guidance unit on top of the Fan Song unit, and a downlink signal from the missile. This greatly reduced the warning time available to U.S. aircraft that relied on the RHAWS, because that system was not built to detect this type of signal. Additionally, the Fan Song system was improved so that it could track aircraft at lower altitudes.[95]

Furthermore, the SAM operators had learned from Operation Rolling Thunder and refined their tactics, reducing radar transmission time. The NVN continued to use decoy SAM sites, and to send decoy transmissions to trigger ARM launches. After the Wild Weasels launched their ARMs at the fake signal, the real SAM site would engage the U.S. aircraft.[96] The NVN's air defense network also employed more varied sensors to assist the SA-2 batteries, which were not usually associated with that system. Chinese AAA radar helped provide range and altitude information accurate enough for SA-2 operators to salvo their missiles in such a way that they would rely upon the warhead's proximity fuse to hit the target, rather than on the Fan Song's guidance.[97] U.S. aircraft observed MiGs flying formation with B-52s and supplying altitude information to SA-2 sites.[98]

U.S. countermeasures during the Linebacker operations resembled those of Rolling Thunder, but were more advanced, reflecting the time that had passed since the cessation of bombing from 1968 to 1972. The USAF also committed more resources toward electronic countermeasures and the Wild Weasel/Iron Hand mission.

Self-protection jamming pods saw action in great numbers in 1972. The USAF used the same pods as those deployed during Rolling Thunder, with the addition of some more advanced models, the ALQ-119 and the ALQ-101. Fighter-bombers initially experienced difficulty with the ALQ-119, because its jamming signal interfered with RHAWS gear, and with the television screens when they used TV-guided bombs. Field modifications and corrections to their assembly by the manufacturer soon addressed these problems. Additionally, rigid pod formations, which were necessary for an adequate pattern of jamming, made the aircraft vulnerable to MiG interceptors. New formations that ensured pilots could "check six" (monitor for MiGs approaching from the rear) and maintain the jamming pattern were necessary, and were put in place soon after the campaign began. These formations involved two two-plane elements weaving back and forth across each other's flight paths to provide cover for each other. This sacrificed bombing accuracy, however, so the more rigid formations were then adopted as the aircraft closed in on the target.[99]

The USAF used chaff extensively during Operation Linebacker. At first, standard practice was to use chaff "corridors." F-4 aircraft—the primary airframe used to dispense chaff barrages in 1972—carried special chaff dispensers and bombs; they would lay chaff patterns on either side of the fighter-bombers' flight paths. These patterns would form barriers between air defense radars and the strike. It took some time for these patterns to develop, so the chaff flights would precede the strike by at least fifteen minutes.[100]

The chaff corridor had four main weaknesses. First, chaff consisted simply of strips of metal foil, so the chaff clouds were very susceptible to wind. Second, the corridor itself pointed like an arrow toward the target, aiding the NVN in its defenses. Third, because chaff aircraft had to precede the strike by such a lengthy period, they were vulnerable to enemy interceptors. Fourth, an F-4 with a full load of chaff was slow and unmaneuverable. The USAF corrected these problems midway through Linebacker I by switching to area chaff barrages and fighter escorts for the chaff F-4s.[101]

The F-105D had been retired from frontline service by 1972, but the F-105G—an improved version of the F-105F—was the primary Wild Weasel aircraft in use at that time. The Wild Weasels employed the STARM in greater numbers than were used during Rolling Thunder. The hunter-killer Iron Hand teams now consisted of two F-105Gs with ARMs and two F-4s armed with AIM-7 Sparrows for self-defense and cluster bombs to destroy SAM and AAA sites.[102] As before, the Wild Weasel aircraft would mark the targets for their wingmen. There were plenty of ARMs available, and the Wild Weasels used them with gusto. From April to September 1972, F-105Gs fired 678 Shrikes and 230 STARMs for three confirmed radar kills and 96 "possibles."[103]

Despite the heavy use of these countermeasures by the USAF, it suffered some losses, most notably twelve B-52s during Linebacker II. Two types of B-52s, the D and G models, took part in Linebacker. Initially, there seemed to be a correlation between the aircraft shot down and its model. The jamming system of the G model was thought to be less able to deal with the SA-2 than that of the D. The more likely cause, however, was connected to bomber tactics and the position of the aircraft lost, both in their formations and in the order in which they proceeded over the target. As historian Charles Hopkins states, "Indications were that bomber tactics were rather stereotyped."[104] As the first cell passed over the target, the NVN would use the information gathered to target subsequent cells. Furthermore, after dropping their bombs, the B-52s executed post-target turns, which slowed them down, changed the pattern of signals emanating from their jammers, and made them easier targets. The challenges of planning such a complicated operation involving dozens of strike aircraft, and potentially hundreds in support, seem to be the source of these "stereotypical" operations.[105]

During the Christmas Bombings, the first B-52 sorties were accomplished by sending the attacking cells to their targets spread out and at the same altitude. After the first three days of Linebacker, the USAF made adjustments that made the NVN's targeting problem more difficult: the times over the targets would be shortened, the three-aircraft cells would all fly at different altitudes, and aircrews were allowed to vary their altitudes as needed. Furthermore, chaff would be distributed over general areas, rather than along the routes.[106]

Individual accounts reinforce the notion that at first, USAF tactics and procedures helped NVN air defenses to be more effective. One B-52 EWO's account relates a number of facts. First, the chaff corridors were not very effective because they tended to blow away with the wind. In addition, the planning staffs did not believe the chaff had blown away, maintaining that it was still there despite vehement insistence by aircrews that it did not remain in place long enough to be effective. Second, SAC procedures tended to be dogmatic. The results of experiments conducted at Eglin Air Force Base between B-52s and an SA-2 system were given more weight than the operational experience of aircrews that had flown over North Vietnam. Third, SAC bombers and their countermeasures had been optimized for low-altitude penetration runs against the Soviet Union; however, the Linebacker raids flew at 38,000 feet. Finally, aircrew training was adequate, but did not encourage flexibility. Specific training for the SAM threat was also unrealistic.[107]

Was the U.S. Air Force's Adaptation Successful?

It seems clear that the SA-2 steadily decreased in effectiveness throughout the war. As noted above, the best way to evaluate the USAF adaptations to the SA-2, given the different goals, operational tempos, and targets of the various air campaigns, is to evaluate U.S. aircraft losses. The primary and secondary sources used in this chapter offer a number of different statistics in that regard: "During the last six months of 1965, the enemy fired an average of 20

SAMs per SAM-lost American aircraft. In 1967, the ratio was approximately 55 to one. During the unprecedented 11-day air offensive of December 1972, it was at least 100 to one."[108]

Part of the reason for this decline in SAM effectiveness was the jamming pod. One study asserted that jamming pods had proven to be particularly effective: "Putting these noise jamming pods on each of four aircraft which fly a rather precise, widely spaced formation presents . . . a large, spatially dispersed source of noise" and "denies range and good direction information to the radars," according to William R. Rambo, director of Stanford University's electronics laboratory. Rambo cited some impressive statistics concerning pod effectiveness, telling the Air Force Scientific Advisory Board that before Project Vampyrus—the pod-testing procedure described above—the NVN was averaging one kill for every thirty-five SAMs launched, but afterward, the ratio was one plane downed per sixty launchings. Anecdotal accounts from fighter-bomber pilots support this estimate.[109] USAF adaptations had not completely blunted the NVN's air defenses, but had made it exponentially harder for the SA-2 to down U.S. aircraft.

The Wild Weasel system was effective as well. Rather than being responsible for the outright destruction of SAM systems, however, the Wild Weasel was primarily helpful because of its impact on deterrence. The NVN found ways of employing the SA-2 despite the presence of ARM-equipped USAF aircraft, but the SAM sites had to worry about their survival when these aircraft were in the area.

In an identical period in 1966, 548 missiles downed 15 planes, none of which had self-protection pods, giving a ratio of 37:1. From April to October 1972, though, U.S. aircraft sighted approximately 2,661 SAMs, and they downed 41 pod-equipped planes. This constitutes a ratio of 65 missiles launched for each plane downed. In terms of aircraft lost to SAMs, however, 1967 was a better year than 1972 for the USAF. As the USAF documentation states, "This would seem to indicate that the pod-equipped fighter-bomber was almost twice as safe as the plane that had none. Differences in weather, types and numbers of defensive weapons, the skill of the defenders, and ability to maintain a pod formation combined to subvert statistical analysis."[110]

An examination of statistics associated with AAA losses paints a different picture, especially when one compares them to losses due to SAMs. During Operation Rolling Thunder in 1965–1968, SAMs downed 5, 18, 28, and 3 aircraft respectively. During that same period, losses to AAA were 142, 267, 260, and 252. Even accounting for the fact that some of the AAA's downings occurred over South Vietnam, losses to its ground fire remained fairly steady for three years, while those caused by SAMs steadily fell.[111] This inevitably leads to the conclusion that the USAF did not find a solution to the AAA problem during the Vietnam War. The increase in SAM losses from 1966 to 1967 indicate an increase in the number of SAMs, not in their effectiveness, as the discussion of ratios in the preceding paragraph indicates.

During Operation Linebacker II, the number of aircraft the USAF possessed really began to have more impact because of the compressed timeframes during which the raids took place. As the twelve-day campaign progressed, B-52 losses decreased. This reflected not only quick tactical adjustments on the part of the SAC units involved, but also multiple raids approaching NVN targets from different directions almost simultaneously, which soon drained the NVN's defensive resources. The SAM batteries simply ran out of missiles.[112] In any case, the interactions between SAMs and AAA that existed during Rolling Thunder, in which the SAMs drove U.S. aircraft into the teeth of the AAA, continued during the Linebacker operations. Most U.S. aircraft lost during the latter operations were shot down by AAA.[113] It seems clear that the USAF adaptation was at least partly effective.

The U.S. Air Force's Organizational Ability

Now that I have described the innovation, the USAF adaptations to it, and the relative effectiveness of each, we can now turn to the USAF's ability to organize to meet the new threat. The USAF was strong in resources, regarding both numbers of aircraft and levels of technical expertise. It was, however, weak in the arrangements of its subunits. The USAF started the war with a shunned adaptation requirement. This weakness led to the misallocation of resources prior to the Vietnam War, leaving the organization ill-equipped to fight the war it was called upon to fight.

Resources

The USAF was strong in resources for a number of reasons. It had large numbers of aircraft to commit to operations in Vietnam. Additionally, it did not lack for trained pilots or other aircrew. Besides these rather obvious indicators of resources, the development of USAF countermeasures also provided ample evidence.

The F-105 is a good indicator of USAF strength in resources. On one hand, the F-105 was not the ideal aircraft for Operation Rolling Thunder. Its emphasis on speed and delivery of nuclear weapons made it less than a perfect fit for the Vietnam War's mission requirements. Its designers had not anticipated the need to deliver conventional weapons against pinpoint targets that were heavily defended with SAMs and AAA. Its lack of maneuverability made it vulnerable to enemy defenses, and a lucky shot could make the aircraft uncontrollable. Even though the USAF knew of the NVN's potential threat in this regard, it undertook no countermeasures until it had lost the first planes.

On the other hand, the USAF's deployment of the F-105 demonstrates that the organization had plenty of resources. First, despite the very heavy casualty rate noted above, the USAF was able to maintain two full-strength operational fighter wings in Thailand for the entirety of Rolling Thunder. There seems to be no evidence of a shortage of F-105Ds (the single-seat version), even though the majority of the airframes ended up as wreckage in Southeast Asia, or were used as spare parts for other aircraft. Second, its speed, payload, size, and rugged construction—especially after the hydraulic system flaw was corrected—made it an ideal candidate for a Wild Weasel airframe. Additionally, the F-105G was still the primary Wild Weasel aircraft even after the F-105D had been retired from frontline service. Furthermore, once the USAF had developed RHAWS gear, it was eventually installed on all fighter-bombers. This demonstrates that the organization as a whole did fairly well at shifting resources to where they were needed.

If we consider electronic countermeasures and specialized weapons separately, the picture of USAF resources is a bit more mixed. Jamming pods were quite rare in the first couple of years of Rolling Thunder. Furthermore, the USAF had to get its ARMs (both Shrike and the STARM) by sharing those programs with the USN. To a large degree, the delay is understandable; weapons systems had reached completely new levels of complexity and adjustments were required. The numerous instances of different aircraft systems interfering with one another serves to illustrate this: electronic jamming not only interfered with RHAWS gear, but also, at times, with weapons delivery systems. This complexity meant that once a problem was identified, the time between deciding what to do and actually implementing the decision lengthened.

In terms of personnel, the USAF was fortunate. Not only did it have a group of pilots who were willing to fly Wild Weasel missions, but also there were plenty of EWOs available to operate the complicated systems required. These EWOs were easily transferred from SAC.

Finally, the USAF demonstrated its capability in resources during the Linebacker campaigns of 1972. While the Nixon administration did not commit the full weight of U.S. airpower to the mission, it committed a substantial portion of it. By 1972, the resources for degrading air defenses had increased substantially. There were more Wild Weasel and jamming aircraft committed to each mission, and the USAF used large numbers of chaff-dispensing aircraft to blind enemy radars. The proportion of SEAD aircraft to strike aircraft had greatly increased since the days of Rolling Thunder. U.S. forces simply overwhelmed the NVN's air defenses, especially during Linebacker II.

The Top-Level Review Panel's Response to the Innovation

Relative to the innovation, the organizational arrangement of the USAF was detrimental to adaptation, and accounted for the significant delay between the first detection of SA-2 systems in North Vietnam and the reaction to them.

If one examines the adaptation requirement relative to the organizational arrangement of the USAF immediately after the introduction of the SAM to Southeast Asia, it becomes clear that the NVN's innovation represents a "shunned" requirement. The two main USAF subunits were SAC and TAC. SAC had significant capability to degrade enemy air defenses (see figure 4.1).

This is illustrated by the fact that SAC B-52s had electronic countermeasures already installed; however, those B-52s were only used for bombing North Vietnam during Linebacker in 1972. Their primary role during Rolling Thunder was route interdiction outside of North Vietnam. Furthermore, SAC's countermeasures had been designed to enable the bombers to avoid enemy defenses on their way to targets inside the Soviet Union, not to take those defenses head-on, as was required for Rolling Thunder. TAC, for its part, factored enemy air defenses out of its calculations entirely. When Rolling Thunder began, the development of self-protection pods was essentially a side project, and no specialized aircraft existed for directly attacking SAMs and AAA.

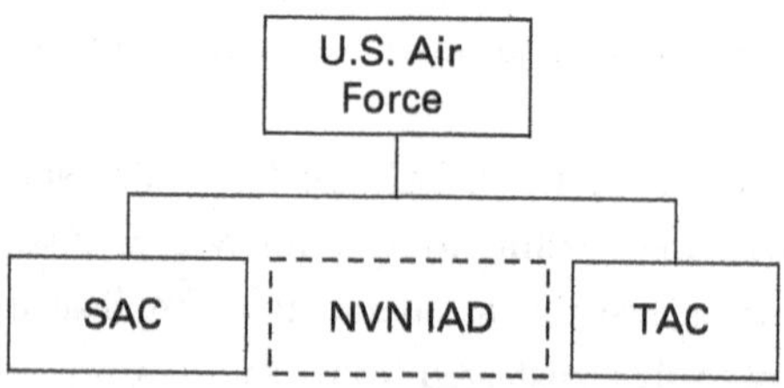

Figure 4.1
The initial adaptation requirement: North Vietnamese integrated air defense innovation relative to U.S. Air Force subunits.

This situation did not remain static, however. The initial losses to the SA-2 prompted the USAF to convene General Dempster's conference, which in turn prompted the development of several technologies and subunits to address this threat. Reconnaissance squadrons, later redesignated "Electronic Warfare" squadrons, of EB-66s started widespread jamming operations in support of Rolling Thunder. The Dempster Commission demonstrates the importance of committing a group of personnel to solve a particular problem. Out of this came the directives to speed up development of jamming pods, develop RHAWS gear, acquire ARMs, and expand the mission requirements for Wild Weasel. This represented the majority of USAF countermeasures used throughout the war. Even though the Dempster Commission was a temporary creation (no evidence was found of its existence after the proposed countermeasures were developed), the function it served as the wellspring of USAF adaptations testifies to its importance. One temporary subunit provided the initial coordinated effort to think about the problem and develop responses. It is also clear that, independent of how one judges the success of Rolling Thunder as a whole, the effectiveness of the SAM in shooting down aircraft eroded over time. While not an ideal organizational response, such as a DOS, the USAF's TLRP represented an organizational improvement. The new relationship between the USAF's organizational arrangement and the adaptation requirement can be seen in figure 4.2.

The SAM was only part of the picture, however. While its effectiveness decreased over time, it still drove U.S. aircraft down to altitudes where AAA

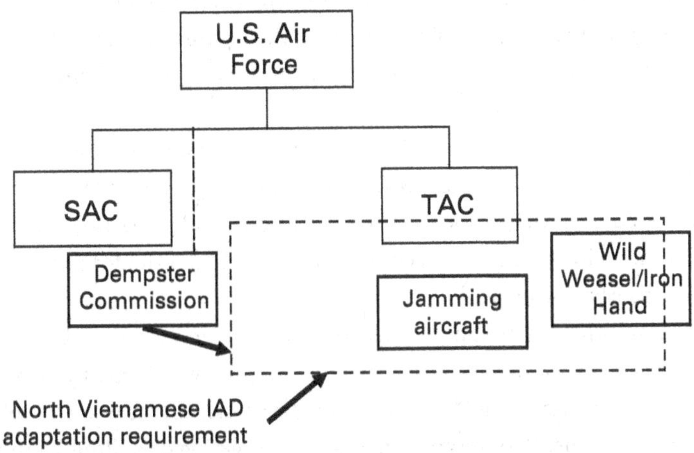

Figure 4.2
The U.S. Air Force's organizational adaptation.

could effectively target them. The USAF never completely solved the AAA problem. Flying high enough to avoid it sacrificed accuracy and made aircraft vulnerable to SAMs. Even denying AAA batteries their radar guidance still allowed them to use barrage fire, because they were placed around the target. U.S. aircraft had to fly past the AAA in order to execute their missions. The only means to completely avoid AAA—flying higher—would not become a better option until the accuracy of weapons delivery improved.[114]

The Wild Weasel's mission required special training and strong motivation of its aircrews. Even though Wild Weasel squadrons did not exist until 1970, a special training pipeline was already in place from the outset of the war. This was important because the mission itself was quite dangerous. Replacement aircrews were always needed, and the veterans who survived could impart their experience to new personnel. The *esprit de corps* that existed among Wild Weasel personnel also helped them face the harsh reality of the dangers of their missions.

If one looks at the entirety of the USAF response, its countermeasures were not completely integrated into one subunit. Jamming aircraft operated in different squadrons from Wild Weasels; however, it was enough to make SAMs less effective.

One of the oft-cited faults of Rolling Thunder—its halting nature—also allowed the USAF time to adapt. The adaptation requirement caught the organization as a whole flat-footed. It had considered the problem before the conflict, but the subunit that executed Rolling Thunder, TAC, had ignored it because of funding and concentrating on other missions. Given the time it took to develop sophisticated countermeasures, the fighter wings in Southeast Asia would have had to absorb most of their punishment from the North Vietnamese innovation before adaptation was possible, had the campaign operated at a higher tempo.

The formation of the TLRP in response to the appearance of the SA-2 clearly points to the origin of USAF adaptation to be from within the service and directed from the top down. We must also consider the alternative theories of civilian intervention and competition as well in order for the case study to be persuasive, however.

The most likely cumulative effect of civilian intervention on USAF adaptation was negative, meaning it hindered adaptation. Civilian intervention in the air campaign over North Vietnam was generally heavy. The Johnson administration directed the goals, tempo, and targeting of Rolling Thunder. In the specific task of adapting to the NVN's IAD, however, civilian intervention was not decisive, and was even counterproductive at times. The first response to the loss of USAF aircraft to an SA-2 was the retaliatory raid that the administration directed; it was a disaster. Furthermore, the nature of the Rolling Thunder campaign as a limited instrument of coercive diplomacy prevented the USAF from dealing with the NVN's IAD in a comprehensive fashion—i.e., unlimited and unduly escalatory from a political point of view. Perhaps the most damaging form of civilian intervention was the Johnson administration's prohibition on bombing NVN airfields that housed MiG aircraft. This did not directly affect the SEAD mission, but it underscores the nature of the civilian intervention in the air campaign: targeting external political actors—namely, signaling the NVN and its Soviet sponsors—rather than targeting internal bureaucratic actors. Civilian leadership was not focused on the adaptation problem.

There is little evidence of competition between subunits of the USAF. Indeed, the primary subunit involved in Rolling Thunder, TAC, was the loser of a long budget competition with SAC prior to the conflict. During the war, TAC did not face any competition from within the USAF regarding the best way to handle the NVN's IAD. As noted in the narrative above, SAC actually contributed significant resources to TAC to assist in the adaptation process. Much of the technology initially available (namely jamming pods), as well as personnel (electronic warfare officers that flew in the Wild Weasels), came from SAC. It should also be noted that the demands and unique conditions of wartime may be able to overcome some destructive competition between bureaucratic actors. While influence and money are always important and often cause conflict between subunits, the demands of accomplishing the mission and saving lives may outweigh them. In other words, bureaucratic barriers exist and cause delays and inefficiencies in adaptation, but individuals and small groups overcome them because the exigencies of a crisis demand it. Further investigation into the factors that lead to greater or lesser bureaucratic competition—especially participation in violent conflict or war—is warranted.

When we broaden the discussion of competition, we must include the USN, the other primary actor in the Rolling Thunder and Linebacker campaigns. Three main factors argue against competition between services as the driver of adaptation. First, both services faced the same political restrictions on their activities during Rolling Thunder.[115] Indeed, the USN faced the same prohibition against striking airfields and hence dealing with the MiG threat. Second, the USN developed both of the ARMs and much of the electronic warning gear used by the USAF, and shared them willingly.[116] The USAF acquired USN systems weeks, and sometimes even mere days, after the

USN introduced them. Third, the route package system that divided North Vietnam into geographic areas also divided USAF and USN responsibilities. Besides striving for superiority in accomplishing their Rolling Thunder missions, inter-service rivalry did not seem to spur successful adaptation to the NVN's IAD. The competition theory implies that the USN would have let the USAF go it alone in developing countermeasures, or resisted the latter "piggy backing" on its programs. These factors seem to mitigate the potential for any effect of competition or inter-service rivalry on successful adaptation.

Conclusion

This case demonstrates that organizational arrangement relative to the adaptation requirement matters. The USAF was well resourced immediately prior to and during the Vietnam War; however, the arrangement of its subunits affected the distribution of resources within the organization and its operational focus. At the outset of the war, TAC was unprepared for the threat of the NVN's IAD. It had to develop an appropriate response in an *ad hoc* manner. The adaptation was largely successful, but there were inefficiencies. Furthermore, the Dempster Commission, while temporary in nature, directed wide-ranging countermeasures and exerted considerable influence on the USAF and its supporting industrial base. With the technology of the time, however, it could not completely resolve the conflict between flying above AAA coverage and bombing accuracy.

5

The British Royal Navy and the Falklands War

In April 1982, Argentina invaded the Falkland Islands. Despite diplomatic efforts by the United States, Peru, and other nations, Great Britain and Argentina became embroiled in a short, high-intensity war, a large part of which involved the naval and air forces of both countries. This case sheds light on the innovator-adapter relationship in a number of important ways.

This case enriches the project in terms of duration, power relationships, and technological determinism. First, as noted in the introduction to this book, sustained naval combat is increasingly rare, and students of naval warfare would be remiss if they did not at least consider its most modern example. Second, the other cases in this volume examine wars of long duration. Analysis of a "short war" adds analytical depth to this examination. This allows the project to examine the effect of strict time constraints on adaptation. Certainly experience and logic tell us that adaptation takes time. Do time constraints preclude effective adaptation altogether?

Furthermore, the Falklands War represents another dimension in the innovator-adapter relationship—the size of the opponents. The cases from World War I and World War II involve Great Powers on both sides of the equations. The Vietnam case has a superpower (the United States) and a small power (North Vietnam) on opposing sides. For the Falklands case, a small power opposed a great (but not "super") power. Finally, as the most "high-tech" naval war in history to that point, this case can help to ascertain how technology affects the ability of innovators and adapters to both innovate and adapt.

This case study reveals that, as expected, short, technologically-driven warfare leaves little time for large-scale reorganization. Adaptation is accomplished, if at all, in the field as the fight progresses. Therefore, in short wars, organizational arrangement and flexibility prior to war takes on more significance. High-technology warfare is a "come as you are" affair. The lead-time in developing complex weapon systems is too long to affect units deployed in the field more than at the margins. It is even more harmful, as this case demonstrates, for a military organization to increasingly specialize for one mission at the expense of others. As vital as the North Atlantic Treaty Organization (NATO) anti-submarine warfare mission was for Great Britain at the time, it still had overseas territories, and one of them—the Falkland Islands—potentially needed defending. If the crisis had occurred just half a year later, the British Royal Navy (BRN) could not have managed to transit to the South Atlantic, let alone prosecute a war there.

When faced with the innovative use of Argentinian air power, the BRN suffered from critical gaps in its ability to organize. It hastily assembled a task force whose combat power was concentrated in two light carriers and their escorts of various ages and capabilities.[1] The ships that the BRN possessed were most suited for anti-submarine warfare (ASW) operations on NATO's northern flank under the cover of shore-based aircraft. It lacked an organic Airborne Early Warning (AEW) capability and a Close-In Weapon System (CIWS), both of which were vital to defend ships against the threat of low-level missiles and aircraft.[2] The BRN's new concentration on the NATO mission reflected the British retreat from "East of Suez" missions in general. The national leadership viewed traditional power projection as unnecessary and costly; thus, in 1982, the BRN was shedding from its inventory most of the ships that gave it that capability (e.g., aircraft carriers and amphibious warfare ships). The transition to a strictly NATO-oriented force was not yet complete, however. The BRN retained or regenerated enough power-projection capability to give Operation Corporate—the name of Great Britain's operation to re-take the Falkland Islands—a chance to work. As unprepared as it was for the type of conflict it faced, the BRN was still a Great Power navy, which was reflected in the size of the task force, its personnel and their expertise, and its logistical capabilities. This combination of over-specialization and a large amount of available resources resulted in victory, but one that was costly in both ships and lives.

The BRN's mission was to establish control of the air and sea around the islands in order to accomplish an amphibious assault.[3] In order to analyze the effectiveness of the Argentinian mode of operation, I will first describe its most important aspects. It is then necessary to examine the state of the BRN at the time. This is because the Argentinian mode of operation was innovative not because it was completely novel—it was not—but because the BRN was in the process of evolving in a way that made it vulnerable to the type of operation that the Argentines carried out. Finally, an analysis of the major elements of the BRN's air and naval campaign will make a verdict on the effectiveness of the innovation and adaptation possible.

The Innovation: Low-Altitude Air Attack

The Argentinian mode of operation against the BRN consisted of three elements: air-launched Exocets; land-based fighter aircraft with gravity bombs and rockets; and a surface fleet that featured a large number of Exocets and protected a World War II–vintage carrier and light cruiser. It was intended to deny the BRN control of the sea around the islands and hence disrupt any attempt to retake them. I will discuss each of these elements in turn. This strategy was innovative because its target, the BRN, was not structured or equipped to counter it at the outset of hostilities.

The newest and most famous component of the Argentinian effort to defend this new territory was the Exocet missile. A medium-range, sea-skimming anti-ship cruise missile (ASCM), the Exocet received targeting data from its launching aircraft, relayed from a scout or the launching aircraft itself. Once launched, it would use inertial guidance until, at a preset point along its flight path, it would turn on its onboard radar. Using this radar, it would home in on the center of the nearest target. Traveling at close to Mach 1 very close to the wave tops, it was very difficult to detect, let alone hit. At a total weight of over 1,000 pounds, the impact of the missile and the detonation of its warhead (and any fuel left onboard) would cause massive damage to the target.[4] Most of the major Argentinian surface units deployed the Exocet, but it was the air-launched version that would give the BRN the most trouble. The Argentinian Navy had only acquired five missiles for their Super Entendard aircraft.[5] Both aircraft and missile are French-made, and Great Britain was able to persuade its ally to stop providing missiles and spare parts after the invasion of the islands.[6] The Argentines pinned much of their hope for repelling the BRN on the Entendard-Exocet combination, expecting to achieve hits on one or both of the British carriers. Without the carriers, the BRN would lose any hope of protecting its fleet from the air, and Operation Corporate would fail.

The second component of the Argentinian effort was its major fleet units. During the war, these were separated into two groups, one based upon the carrier *25 de Mayo*, the other centered upon the cruiser *General Belgrano*. The carrier's air wing consisted of a small number of A-4Q Skyhawk attack aircraft, supported by surveillance aircraft and helicopters. The Skyhawks did not carry the Exocet. Rather, they would use a more traditional mode of attacking shipping: bombs, guns, and unguided rockets. The naval aviators were the only ones in the Argentinian military trained to attack naval targets.[7] In the short time between the invasion and the arrival of the British task force, they imparted some of their knowledge to the pilots of the air force. The Argentinian Navy, which possessed British-built Type 42 destroyers (described in more detail below), used its own vessels as practice targets to get its own and the air force's pilots up to speed. These practice runs helped Argentina discover the Type 42's weaknesses, and confirmed the need for low-level attacks to get beneath the BRN radar and missile coverage; however, high casualty rates were still expected.[8] The naval pilots also knew that special bombs were necessary for very low-level attacks on British ships. The type of bombs required were "retarded" bombs, called "snake eye" in the U.S. inventory. Bombs dropped at very low altitude threaten to hit the delivery aircraft with shrapnel, so these bombs deploy air brakes to slow them down. This allows the delivery aircraft to escape their blast radius before the bomb explodes. It also has the effect of giving the bomb's fuse time to arm. The Argentinian Navy's A-4's used these bombs in the air campaign, but the air force did not.[9] Given that these bombs were in the Argentinian inventory and that the Navy imparted at least some knowledge to the air force on how to bomb ships, it would be expected that the air force would have used these bombs as well. The number of "duds" that hit British ships is evidence that it did not. Furthermore, the experience of the first few missions posed a stark conundrum: in order to survive British missiles, guns, and their Sea Harriers (SHAR), the Argentines would have to fly below fifty feet in the combat zone. Below fifty feet, however, their bombs would not explode.[10] The Argentines did not correct the fuse problem or develop new tactics to address it.

The surface units of the fleet, most notably the *General Belgrano*, posed a significant threat to the British task force as well. While the air-launched Exocet arsenal consisted of a mere five missiles, the destroyers and corvettes of the Navy were capable of launching thirty-six.[11] The cruiser, originally the light cruiser USS *Phoenix* of the *Brooklyn* class, mounted nine 6-inch and eight 5-inch guns. Also, it was the one ship in either the Argentinian Navy or the BRN that possessed armor protection. Together with the Exocet-armed smaller vessels, Argentinian use of this equipment posed a serious threat to the British task force.

In terms of numbers, the most significant component of the Argentinian effort was its air force. Its inventory consisted of an array of foreign-produced aircraft. These included the U.S.-built A-4, the French-built Mirage, and the Israeli Dagger. The Dagger lacked adequate navigational equipment for the trip to the islands, so the air force used Learjets, which had the necessary avionics and navigational equipment, as pathfinders for them. The Dagger pilots were on their own on the return trip, using only a chart and dead reckoning to get themselves home.[12] To put the demands of this transit distance in perspective, only U.S. Air Force fighter aircraft in the Vietnam War faced longer transit distances, but were much heavier planes that were supported by fleets of tankers. The A-4s of the Argentinian Air Force had midair refueling capability, but only two tankers were available. The time over target for all types was severely limited because the transit distance alone was between 380 and 420 nautical miles. On the islands themselves, the runway at Port Stanley could not be lengthened as needed to accommodate fighter aircraft in the time available, so Argentinian aircraft had to fly from mainland bases. Even though the British line of communication stretched for thousands of miles, it was Argentina that suffered more from the effects of distance.[13] Dive-bombing would have put the aircraft in the best engagement envelope of British anti-aircraft weapons, because the bombing runs would have started at medium or high altitudes. The only viable tactic against the BRN ships was the low-level bombing run, using 500- and 1,000-pound unguided bombs. Unlike the navy, with its five guided air-to-surface Exocets, the air force did not possess any guided weapons. Furthermore, it had inadequate numbers of reconnaissance aircraft, severely limiting its ability to gather tactical information.[14]

The importance of the difference between the air force and navy becomes readily apparent when one looks at the number of sorties flown compared to the number of losses they inflicted on the British fleet. Even though the Argentinian Navy flew only 23 of 495 total sorties, naval aircraft accounted for three of the six sunken British ships.[15]

The British Royal Navy in 1982

The Argentinian mode of operation was not strictly innovative; however, the state of the BRN at the time made it so. This was because British naval policy was in the final stages of a transition from the "East of Suez" naval policies to the "Eastland navy," which encompassed the seas in the NATO area of operations. The "East of Suez" policy was one in which Great Britain would maintain naval forces capable of intervening in the Middle East, Africa, and Asia. Decolonization meant that these areas faced increased likelihood of instability. Even though the United States had supplanted Great Britain as the preeminent global naval power, the U.S. Navy was unwilling to extend its presence into areas such as the Indian Ocean. Great Britain reasoned that the nuclear balance in Europe made war unlikely there, and that Soviet attention would likely be drawn to "soft spots" in the developing world that were ripe for communist agitation. On this basis, the BRN was able to sell its expeditionary role to the British government in the late 1950s.[16]

A combination of inter-service rivalry with the Royal Air Force (RAF), budgetary austerity brought about by economic woes, and the changing world situation caused the BRN to lose this role in the subsequent decades. The RAF was particularly hostile to naval air power. The two services had an adversarial relationship dating back to the RAF's gaining control over the fleet air arm in the last year of World War I. The RAF had been able to focus on strategic bombing partially at the expense of the BRN's budget during the interwar years.[17] The competition between the two services was just as intense during the budgetary austerity that accompanied the contraction of the British Empire. The RAF was able to use the cost of maintaining carriers as well as their perceived vulnerability to land-based aviation to lobby against the maintenance of a carrier fleet. The Air Ministry allied itself with the Treasury, feeding the latter with anti-carrier arguments. These organizations argued that a carrier fleet was expensive, vulnerable, and redundant, because the RAF could provide air cover to the fleet. Furthermore, with island bases, the RAF's transports could compete with the BRN's amphibious capability. This culminated in the cancellation of CVA 01, the proposed new carrier, in the mid-1960s, and the scrapping of the BRN's last carrier, the HMS *Ark Royal*, in 1978.[18] These ships were vital to protecting expeditionary forces from land-based air forces—Soviet bombers armed with long-range anti-ship missiles, for example—so when they disappeared, the wider East of Suez mission came into question as well.

British defense policy in the early 1980s reflected this transition. By 1981, the Ministry of Defense envisioned six roles for the BRN: strategic nuclear deterrence, containment of Soviet naval forces, defense of reinforcement and supply shipping, ASW support for NATO, deployment of amphibious forces to NATO's northern flank in Norway, and protection of NATO and British merchant shipping. The RAF would provide air surveillance and defense from its shore bases.[19] The BRN was to be turned into a predominantly ASW force.[20] These missions were entirely dissimilar to what Operation Corporate required of the BRN.[21] Any NATO amphibious operations in Norway were to be conducted at the end of a short supply line, with ample land-based air cover, and within close proximity of allied NATO forces, namely U.S. aircraft carriers.

With the loss of the *Ark Royal*, the BRN also lost its long-range interceptor (the F-4 Phantom), strike aircraft (the Buccaneer), and its only airborne early warning (AEW) aircraft (the Gannet).[22] The BRN possessed an ASW carrier (the *Invincible*) and a commando carrier (the *Hermes*). The former carried numerous helicopters for the ASW mission, while the latter embarked troops and helicopters for amphibious assault. By 1982, fixed-wing aircraft had returned in limited numbers with the introduction of the Sea Harrier, but the BRN still lacked an AEW capability.[23] It was these two small carriers, with their augmented wings of the SHAR, which would face the Argentinian air attacks.[24]

Invasion and Response

Argentina's ruling junta had good reasons to believe that its move to occupy the Falklands would not be opposed in any meaningful way. Argentina was strategically much closer to the islands, and the BRN's line of communication extended more than 8,000 miles. The United Kingdom was also committed to NATO, and the Argentinian leadership doubted that it would strip enough assets from this alliance to retake the islands. The junta calculated that a serious British reaction was "scarcely possible" or "totally improbable."[25] While the timing of the Argentinian invasion was questionable from a strategic standpoint, the ruling junta had strong domestic political motivation. Internal dissent was on the rise, and economic woes demanded painful fiscal action by the government, which calculated that an external war would provide the necessary internal unity to enable it to undertake and enforce its potentially unpopular economic reforms.[26]

The Argentinian Navy and its marines were responsible for the landing operation. The 2nd Marine Battalion, consisting of approximately six hundred personnel, had the support of about sixty commandos. The marines landed in amphibious armored personnel carriers (U.S.-made "amtraks") and used the island's road network to reach Port Stanley from the beaches. Meanwhile, the commandos attempted to storm Government House, the islands' capitol building. Once obstacles were cleared from the air strip at Port Stanley, air force C-130 transports ferried in 1,000 soldiers from the army. This force consisted of soldiers conscripted from across Argentina, and was primarily symbolic.

To oppose the landing force, the islands only had seventy-nine British Royal Marines and sailors. In the short combat that followed, the British garrison was overwhelmed but suffered no casualties, while Argentina incurred some wounded and a handful of killed in action. The Argentines later replaced their relatively elite marines and commandos with more army units, which were predominantly made up of conscripts and led by professional non-commissioned and commissioned officers.[27]

Once the islands were secure, the Argentinian Navy formed three task groups in preparation for the BRN's arrival. Task Groups 79.1 and 79.2 consisted of the carrier *25 de Mayo* and her escorts, plus two oilers. These groups sailed 150 miles north of the Falklands. The cruiser *General Belgrano*, two destroyers, and an oiler made up Task Group 79.3, which sailed southwest of the islands.[28] The air force gathered its available fighter aircraft at land bases that were closest to the islands. Argentina did not lengthen the small airstrip at Port Stanley in order to accommodate fighter aircraft. The logistical demands of the invasion and occupation, coupled with limited Argentinian support capacity, prevented them from improving the airfield; accomplishing these two tasks simultaneously was not possible. This would have important consequences for both sides, as it greatly lengthened the round trip of at least 760 nautical miles for Argentinian land-based aircraft, and allowed the BRN to get much closer to the islands than it otherwise would have.[29]

The British Royal Navy and Operation Corporate

The BRN required ground forces to take back the Falklands, but naval forces were needed to transport them there and protect them before, during, and after any amphibious landing. There were no suitable land bases anywhere near the islands. Therefore, the entire effort for this operation depended upon the fleet, especially its carriers.[30] The BRN assembled a task force that eventually amounted to thirty-five warships, supported by dozens of auxiliary ships and merchant vessels converted for naval use. The only air cover available came from the SHAR of the two light carriers in the British inventory.

Even though it was in the throes of a steady but increasingly precipitous decline from its position of dominance in the first half of the twentieth century, the BRN enjoyed the confidence of its nation. As David George Boyce wrote, "The Navy is one of Britain's prime assets . . . besides its intrinsic worth, the Navy is a hardy and vivid fibre in the cloth of the history and traditions of the nation."[31] The time of "Britannia rules the waves" was long over, but confidence in the Navy, both from within and without, was still strong. Budgetary pressures had not diminished the attitude that the BRN was still a Great Power navy, at least in spirit. These factors were important for two reasons. First, the national leadership had some confidence that the Navy could meet the difficult challenge in the Falklands with some expectation of success. Second, the BRN leadership was willing, and in fact eager, to demonstrate that the organization could accomplish a task that was clearly maritime in nature and that it was relevant to contemporary defense policy.[32] The junta's attempt to present Great Britain with a *fait accompli* had backfired.

An array of escorts supported the two BRN carriers. The Type 42 destroyers had a strong area-defense weapon in the Sea Dart, but lacked a point-defense system.[33] The inability to engage small aircraft at short range was a major liability, but "was not considered of great importance at the time [of the classes design], given the nature of the threat in Eastland waters and the expected performance of the Sea Dart missile."[34] The Type 22 frigate was large, with good sea-keeping qualities and an excellent point-defense missile (the Sea Wolf), but only two were in service. The task force later received among its reinforcements a frigate of the older, *Leander* design, which was upgraded with Sea Wolf. This ship, the HMS *Andromeda*, arrived in the middle of the operation but provided the carriers with an extra "goalkeeper," or last line of defense.[35] The Type 21 frigate was primarily equipped for ASW and mounted the inferior Sea Cat short-range anti-aircraft missile.[36] Despite many of these ships carrying Exocet ASMs, the result of this combination of small carriers and ASW escorts meant that the main striking power of the BRN lay with its nuclear attack submarines (SSNs).[37] With their stealth and load of heavy torpedoes, the SSNs were the only units truly capable of hitting enemy surface vessels decisively at a long distance from the task force. The BRN, however, faced the difficult task of having to fight through an enemy fleet with significant surface assets while operating far from any friendly land base. As Alastair Finlan stated, "The Falklands campaign required all the capabilities that the defense review tried to marginalize or withdraw."[38]

The weapons that the BRN took to sea in 1982 presented a mixed bag as well. Sea Dart was an excellent area-defense weapon, but the missile could not engage sea-skimming targets.[39] The Sea Wolf could engage those types of targets. It was a new system, however, and, as it turned out, could be confused by targets equidistant from its platform.[40] More of the BRN's ships had the older Seacat missile. It was designed to engage low-flying targets, but was much less capable than Sea Wolf and could not engage Exocet. Concerning gun armaments, the primary naval gun, the 4.5-inch, proved to be excellent for shore bombardment. It was not able to track fast-moving aerial targets at close range, though. Smaller caliber weapons, such as 20-millimeter autocannons and general-purpose machine guns, were more suited to engaging targets too small for the big guns, but did not have the fire control to hit fast-moving aircraft reliably.[41]

The SHAR was the BRN's last vestige of carrier-borne fixed-wing aviation. Its primary mission was to provide air support to the ASW forces committed to NATO. It was designed to defend convoys against shadowing Soviet aircraft; provide air reconnaissance against the Soviet fleet; and strike that fleet with nuclear weapons if the situation required. The fleet was augmented with the RAF's version, the GR.3, which lacked the BRN version's radar, avionics for shipboard operation, and anti-corrosion fittings.[42] Both the BRN and the RAF versions received improvements before they went into action in the South Atlantic. The GR.3s needed tie-down fittings to keep the aircraft in place on a pitching carrier deck, reinforced landing gear, anti-corrosion measures, modifications to its transponders, and an inertial navigation system (INS) upgrade.[43] Because of the pressures exerted by a wartime environment, the Ministry of Defense was able to acquire new transponders from defense contractor British Aerospace in just eight days. In eighteen days, the contractor for the INS, Ferranti Corporation, delivered an external platform that provided a workable upgrade for the system.[44]

The United States helped modify British aircraft as well and, most important, it sold Great Britain a new, all-aspect version of the AIM-9 Sidewinder air-to-air missile. The Sidewinder uses infrared radiation to home in on an aircraft's heat signature, primarily engine exhaust. This meant that the missile was most useful when engaging a target from the rear. The new version had a more sensitive seeker that could distinguish heat generated by friction—for example, that generated by contact between the leading edges of an aircraft fuselage and the air. This allowed the updated SHAR to engage enemy aircraft head-on, unlike earlier versions.[45] At first, the aircraft carried defensive chaff and flares, used to decoy radar-guided and heat-seeking missiles, respectively, in an *ad hoc* manner, wedged between the fuselage and the speed brakes or the bomb pylons and the bombs themselves. A U.S. defense contractor eventually provided a dispenser in time for the war.[46] The SHAR, however, did not possess a long-range radar-guided missile like the retired Phantom, nor did it possess the latter's bomb capacity, nor was it supported by an AEW aircraft.

The extra modifications to the SHAR would prove vital to the British war effort. Before the war, these modifications would have taken years to accomplish, but Great Britain completed them in record time out of necessity. On paper, it faced a daunting numerical inferiority to Argentina. The task force brought with it 20 SHAR, which faced 34 light-support Argentinian aircraft on the islands and 110 fighter-bombers from the mainland.[47]

The BRN was in a state of flux when the 1980s began. If the Argentinian invasion had occurred just months later, Operation Corporate would not have been possible. The budgetary environment at the time was austere, and the organization could not support the ships it had. This was because the British government thought its expeditionary mission was too expensive and no longer necessary. The carriers, modest as their air wings were, were essential to the operation's success. In 1982, they and other important ships were on the budgetary chopping block. *Hermes* was nearly at the end of its service life and therefore a prime candidate for disposal. But age was not the only factor. Despite the relative newness of the *Invincible*, the government was even set to cut it as a cost-saving measure.[48] Also, one of the major amphibious units, the *Intrepid*, had been stripped of its stores and made ready for either placement in the reserve or disposal, its crew sent on to new assignments.[49] Two fast replenishment ships, an oiler, and a stores ship were also on their way out of the BRN. The stores ship was in reserve, while the oiler was in transit to Chile to be sold.[50] These ships were vital to the replenishment at sea (RAS), which enabled the task force to transit such large distances and sustain operations. Finally, several of the dockyards and maintenance facilities that would prove to be instrumental in getting the task force ready for sea were slated for closure.[51] When the war started, the BRN was able to reverse these decisions and retain or regain the capabilities these units possessed.

The BRN also boosted its capabilities through Ships Taken Up From Trade (STUFT). These were merchant vessels pressed into service for logistical support. The STUFT force consisted of forty-three ships, which were used as supply ships, troop transports, oilers, aircraft transports, hospital ships, and ocean-going tugs, among other things.[52] Each of these ships required parties of naval personnel and unique modifications to make them suitable for war service, including communications gear and helicopter-landing facilities. The naval dockyards accomplished this difficult task under severe time constraints, and their success amazed even those within the BRN.[53] Under normal circumstances, new weapons and electronics installation takes months to accomplish. The naval dockyards also did considerable work on the BRN's warships within a short amount of time. An *Invincible* crewmember observed that another ship at the Portland dock had, in a four-day period, three new weapons systems installed, two new sensors fitted, four new pieces of communications equipment installed, ammunition and stores embarked, numerous defects "put right," and each watch given one day's leave.[54] This kind of logistical support was vital in getting such a large number of ships ready to fight far away in the southern hemisphere.

The British Royal Navy's Adaptations

The BRN's adaptations came in response to three types of threats: first, the air-launched Exocet; second, low-level bombing; and third, the surface warfare threat. Specific incidents reveal how—and how well—the BRN adapted to each type of threat. The timeline in figure 5.1 puts these incidents in context with each other.

The Argentinian stockpile of Exocet ASMs was limited, but the threat posed by this weapon was very serious. Great Britain had used diplomatic pressure on the Exocet's country of origin, France, to cut off further shipments to Argentina.[55] The BRN, however, had no proven countermeasure in its inventory at the time. As noted above, the launching aircraft and missile could almost completely avoid BRN air defense weaponry. An aircraft could launch an Exocet below the radar horizon of British ships, so the launch often went undetected. Furthermore, the missile flew at such a low altitude that it was detected only minutes or seconds away from impact. Its low altitude also meant that radars, especially missile guidance radars, often lost it against the background of ocean waves. Both sides knew that the *Hermes* and *Invincible* were vital to the BRN's operation, and if either of those ships were put out of action, the odds of the BRN's success would be dramatically reduced. One hit by an Exocet was enough to remove a ship from battle. The fuel and ammunition for the carriers and the fire hazard posed by missile impacts made for a deadly combination. Two incidents, one involving the destroyer *Sheffield*, and the other the merchant vessel *Atlantic Conveyer*, encapsulate both the threat and the BRN adaptation to it. The BRN adaptation to the Exocet consisted of two elements: a layered defense and decoys. Both had their benefits and drawbacks.

- 2 April: Argentinian invasion
- 6–7 April: British corporate task force sails
- 30 April: RAF Vulcan bomber strike on Port Stanley airfield
- 2 May: Argentinian cruiser *General Belgrano* sunk
- 4 May: *HMS Sheffield* sunk by Exocet
- 21 May: British landing operations at San Carlos Water
- 25 May: *HMS Coventry* sunk by bombs
- 25 May: Exocet sets *Atlantic Conveyer* alight, ship later sinks

Figure 5.1
The Falklands War timeline.

Layered Defense

Based on intelligence estimates of the threat, the task force commander, Admiral Sandy Woodward, deployed his ships in a layered defense (see figure 5.2). The outermost layer was the SHAR combat air patrol (CAP), although without AEW aircraft available, they were limited to their own radar, visual sighting, and surface ships' surveillance radar. The BRN was able to equip some of its helicopters with electronic gear that could pick up the Super Entendard's distinctive Agave radar.[56] The picket destroyers formed the next layer with their long-range Sea Dart. A third layer consisted of escorts patrolling in sectors behind the pickets. Inside that layer of escorts was a fourth layer, the ships of the Royal Fleet Auxiliary (RFA). Possessing no significant weaponry but with large radar signatures, these ships would confuse the Exocets' sensors with additional targets. Woodward calculated that the operation could survive the loss of RFA ships, but not the loss of a carrier. The "goalkeeper" frigates were the fifth, innermost ring of defense around the carriers. There were only two Type 22 frigates in the inventory; each shadowed a carrier and would stop any "leakers" with their Sea Wolf missiles, as well as confuse them with additional targets. *Invincible* had its own Sea Dart–missile system, but neither carrier had a CIWS installed. The frigates would make up for the carriers' lack of close-in or low-altitude self-protection weaponry.[57]

The BRN provided the carriers with strong protection. While the Exocet was a guided weapon, its warhead was not smart enough to distinguish

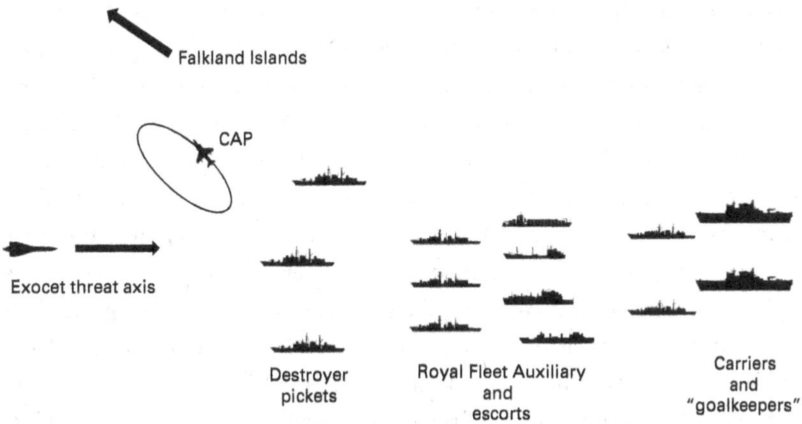

Figure 5.2
Defense in-depth: the layers.

between blips on its radar. The imposition of so many alternative targets between the missile and the carriers was sure to distract the former. This adaptation, however, relied upon Argentina getting insufficient targeting data on the carriers' positions, and left the picket ships extremely vulnerable. The latter, while not as vital as the carriers, were not an infinite resource. Additionally, their missiles and radar, while not capable of engaging low-level targets, still made attacking the BRN a very hazardous endeavor. They were not completely expendable.

The sinking of the *Sheffield* illustrates these drawbacks. In the days leading up to the attack, the BRN ships had been subject to numerous false missile-launch alarms. Virtually everyone in the task force knew how vulnerable their ships were to the Exocet, and anxiety was running high.[58] Guided to the *Sheffield*'s general area by a Neptune targeting and surveillance aircraft, two Super Entendards closed to within Exocet range of the ship.[59] After "popping up" to acquire targets for their missiles, they each launched one missile and turned away at high speed.[60] Despite a brief radar contact from another ship in the screen and electronic signals indicating that an Exocet was in the air, *Sheffield* was caught unaware by the low-flying missile. The Exocet hit *Sheffield* amidship, causing a massive conflagration: "We saw the smoke when *Sheffield* was hit. She was sacrificed to stop the Exocet missile from hitting us, as we were the real target. *Hermes*'s captain, Middleton, told us it would have taken four Exocets to sink one of the carriers, but that one missile would have affected our operational capability for some time."[61]

Even though the warhead failed to detonate, fires caused by the missile's unspent fuel and flammable ship materials quickly raged out of control. Eventually *Sheffield* was abandoned. The other missile failed to acquire a target and crashed into the sea among the third defense layer of escorts.[62] Despite being aware of the threat and having all of the BRN's assets on full alert, a missile had still achieved a hit on an important vessel. The layered defense worked for the carriers; however, it would be very costly in terms of the BRN's most effective anti-air units. More encouraging for the BRN was the poor decision-making of the Argentinian pilots: "No attempt was made to select a high-value target, or even to check the identity of the ship at which they had launched 40 percent of the Argentinian Navy's inventory of this irreplaceable weapon."[63] Argentina had indeed set clear priorities in its targeting. The British carriers were at the top of the list, followed by logistical support ships, ships close to land, and aircraft. Warships were the last priority. Despite this, when the pilots encountered warships first, they attacked them.[64]

The other most effective method to counter the Exocet was to decoy it away from valuable craft by providing false targets. Just as the missile's warhead could not distinguish between ships acquired by its radar, neither could it tell the difference between ships and other large radar-reflecting objects like clouds of chaff. BRN ships were already equipped with chaff before the war. It was deployed by means of canisters fixed to the ship's deck. When triggered, they would fire in a pre-set pattern and the canisters would explode above the ship to form a cloud of metal foil, which would reflect radar signals and provide a false target. The BRN also deployed helicopters with radar-reflective panels hanging from their fuselages above the task force. These had the same effect as chaff, but unlike the clouds of metal foil, they did not disperse and could be repositioned.[65]

The loss of the *Atlantic Conveyer* illustrates how effective and dangerous chaff could be. In an attack profile very similar to that which worked against the *Sheffield*, the Argentinian Navy launched two Exocets at the BRN later on in the conflict, after the first amphibious landings. Upon warning of the attack, the ships of the picket line fired chaff, which decoyed the missiles away from the warships. One missile emerged from the chaff cloud, started searching for another target, acquired the *Atlantic Conveyer*, and struck home. Again the warhead failed to detonate, but caused a massive fire; the ship, filled with important supplies and spare helicopters for the land campaign, had to be abandoned.[66] The ship's cargo of helicopters would have enabled moving an entire brigade from San Carlos to Port Stanley—the capital and ultimate objective of the campaign—by air, greatly reducing the time that the ground campaign would have taken.[67] Chaff quite effectively protected the warships, but had decoyed a missile into an important fleet asset. The loss of the transport helicopters and other supplies onboard made the land campaign logistically more difficult.

Argentina's low-level bombing campaign presented the BRN with different challenges. The Argentinian aircraft had to fly over their targets, making it possible to engage them. Unlike the Exocet, however, shortages of bombs were not an issue for Argentina during the campaign. Sea Dart, the primary anti-air weapon of the BRN's fleet, was very effective, but the Argentinian pilots simply flew low enough so that this missile had few opportunities to engage them.[68] The BRN adapted to this more basic threat by using intelligence-gathering assets as proxies for AEW and interposing picket ships between the direction of the attack and its more vulnerable ships. Operations in support of the amphibious landings best exemplify these adaptations, as well as their strengths and weaknesses.

Without AEW, the BRN resorted to intelligence as one means to counter the Argentinian threat. It positioned Special Air Service (SAS) commandos near Argentinian air bases; they would then radio the fleet when they observed attack aircraft taking off on missions.[69] In addition, as these bases were near the ocean, SSNs performed the same mission from periscope depth. This allowed the BRN to use its limited complement of SHAR more efficiently, as it could wait to launch or reinforce the CAP when it was more certain an attack was on the way. The SHAR would be ready to pounce when the Argentinian aircraft arrived. Even if they were not able to stop the attacks, shooting down the retreating bombers helped attrite Argentina's limited pool of aircraft and pilots, and lowered the survivors' morale. The volume of attacks soon increased to a point where radio warnings were of limited utility, however.[70]

The use of picket ships to counter attack aircraft was similar to that used against the Exocet. The chain of command and division of responsibilities were fairly clear; however, this arrangement was not always followed. Organizationally, the captain of the *Invincible*—Captain Jeremy Black, a naval aviator—was in charge of air warfare, including air defense. The task force commander—Woodward, a submariner by trade—exercised his authority in this arena as well, however. The senior SHAR pilot in the operation, Commander Ward, asserts that Woodward positioned the picket ships too far out for them to provide adequate protection. He also often positioned the CAP too high for it to do anything more than punish the attackers after they had already damaged or sunk BRN ships:

> During the defense of San Carlos, *Hermes*'s Sea Harrier team were told by the admiral's staff to stay up at high level over San Carlos Water area, but not to attack the enemy aircraft until after they'd attacked their targets. Then, as the admiral later wrote in his book, they were to "swoop down like hawks on the tails of the enemy and shoot them down," closing the door after the horses had bolted, in the worst possible way.[71]

A more competent positioning of the CAP might have enabled the BRN to both attrite the attackers and protect its ships. Furthermore, Woodward often shifted the *Hermes* SHAR to bombing missions without Black's knowledge.

An unrelated but telling incident demonstrated Woodward's lack of understanding of air warfare. In the longest bombing mission ever up to that point in the operation, a single RAF Vulcan bomber took off from Great Britain and, after several midair refuelings, dropped a stick of bombs on the Port

Stanley runway. Only one bomb struck the airstrip, and subsequently, it was quickly repaired. While this flight was a stupendous logistical achievement, it was completely unnecessary because the SHAR could have accomplished the same mission more effectively with fewer resources. Furthermore, Woodward yielded to RAF demands that the ships of the task force stay "weapons tight" while its bombers were in the area in order to avoid a friendly-fire incident. The conditions under which the BRN could protect itself were extremely limited under these rules of engagement. Given the potential speed at which modern naval engagements occur, this was not entirely advisable for a fleet in a war zone.[72] Most fleet commanders would view such a risk as unacceptable.

Inter- and intra-service rivalries plagued both the British and Argentinian war efforts, and the friction between Woodward and his air warfare commander could have its origins there.[73] But if one sets aside the inevitable friction between warfare specialties—in this case submariners and aviators—the lack of AEW aircraft meant that the CAP and picket ships could only be so effective, regardless of who controlled them or how expertly they were used.

Layered defense has been the established method of anti-aircraft defense of a naval task force since World War II, and despite having had inadequate weapons with which to counter the low-level threat, layered defense was still the most effective adaptation during the Falklands War.[74] This did not mean, however, that it was popular with the operation's officers and sailors. Chief Petty Officer Arthur Gould of the frigate HMS *Arrow* bitterly noted during the campaign, "Our Admiral, safe out of the line of fire on the *Hermes*, seems to be trying to get all of the existing frigates and destroyers sunk or damaged, and he is succeeding."[75] Woodward undoubtedly knew about the grumbling within the ranks, but was steadfast: "Up until now . . . you have seen fit to take the Queen's shilling. Now you must stand by to front up and earn it the hard way."[76]

From Gould's position, Woodward's actions would indeed have seemed perplexing; however, the heavy losses suffered by the British picket ships were due to the fact that first, they were indeed expendable and the carriers—especially *Hermes*, given that it carried both half the SHAR and much of the landing force—were not; and second, the pickets had not been designed for the role they were forced to fulfill in this type of defense. Their radar capability was not good enough, they were not designed to be survivable, and they lacked fully integrated automatic fire control for their weapon systems.[77] Such a combination of the ships' capabilities, or lack thereof, and the tasks demanded of them almost guaranteed losses. As Captain David Hart Dyke wrote after the war, "With the ships and weapons at our disposal, designed for a very different war, we were always going to be doing it the hard way. (That, unfortunately has all too often been the way of the British in war.)"[78]

The operation's defense of the landing beaches is most indicative of the BRN's employment of the layered defense. There was a myriad of engagements and small actions between Argentinian aircraft and British ships and aircraft at the landing zone, but the summary below of the details they share in common will paint a good picture of the innovator-adapter relationship in this conflict.

Woodward chose as the invasion beachhead an inlet known as San Carlos Water. This body of water is situated on the western side of the eastern Falkland Island, and its mouth opens into Falkland Sound, which bisects the islands. On D-Day, the BRN's vulnerable amphibious warfare ships, troop transports, and RFA ships sailed deep into San Carlos Water to affect the landings. The escorts deployed toward the mouth of the inlet, in the sound, and at the mouth of the sound in order to disrupt Argentinian air attacks. The cliffs and steep hills surrounding San Carlos Water provided shelter: "Once we decided there was no submarine threat, we packed the ships very close together, under the shelter of a bluff of land about eight hundred feet high, which was just high enough to disturb the aircraft as they came in to attack."[79]

At the end of their long transit to the islands, Argentinian aircraft faced a difficult choice: they could use the land as cover, avoiding BRN warships but also severely reducing their time with which to acquire targets, or they could fly through Falkland Sound and into the mouth of San Carlos Water, giving them more time to pick targets but also flying over the escorts. After the British landings were well underway, the army deployed its Rapier missile system on the hilltops surrounding the inlet. Designed to engage low-flying aircraft, these missiles added an extra hazard for Argentina.[80] The escorts with short-range guns and missiles occupied the more confined waters of the sound. Destroyers equipped with Sea Dart sailed in the open waters at the mouth of Falkland Sound, usually escorted by one of the Type 22 frigates with Sea Wolf. The result of this arrangement of units was that the attacking aircraft had to brave virtually every anti-air weapon that the BRN possessed if they used the over-water route (see figure 5.3).

Many Argentinian sorties failed because they were chased away by the CAP that formed the BRN's outer layer of defense. Very significantly, the Argentinian pilots flew so low that many of their bombs did not have time to arm themselves before they hit their targets. They failed to correct this during the conflict, and many BRN ships survived because they were hit by "duds."[81]

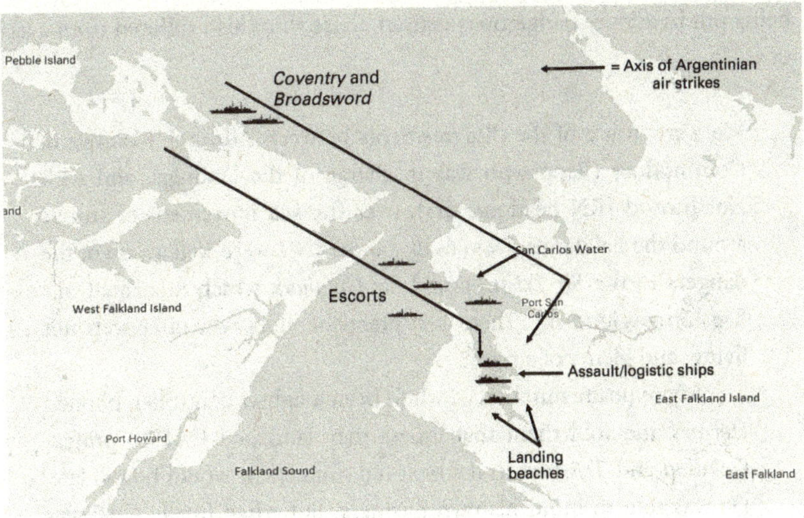

Figure 5.3
Defense in-depth: San Carlos Water.

Furthermore, Argentinian pilots failed to attack the vital amphibious warfare ships and instead attacked the escorts, in part because these were the targets they first encountered and thus were able to hit them without being shot down.[82]

The BRN landings at San Carlos Water on May 21 represented the start of the most testing period for the British operation. Just as the Argentinian pilots' acquisition time was curtailed by the overland route, so too were the British ships' chances of reacting to an attack from that direction. It was a harrowing time for the men: "We were attacked by bombs which skipped along the water like ducks and drakes, some of them bouncing over the superstructure and one going between the masts."[83] By the end of the campaign, only two of its frigates remained unscathed, and when the *Coventry* went down, all of the operation's original picket ships had been sunk.[84] The lion's share of the BRN's ship losses during the war occurred around San Carlos Water. As with their Exocet attacks, the Argentinian bombing runs targeted the first ships that the pilots acquired, rather than the most decisive ones—in this case, amphibious warfare and troop transport ships. The only amphibious warfare ship lost by the BRN, the *Sir Galahad*, was sunk while it was outside of the operation's protection; together with another RFA, *Sir Tristram*, it had sailed forward from San Carlos Water to Bluff Cove, on the other side of East Falkland Island, in order to speed up the BRN's advance on land. In addition to being put in a more dangerous position, these ships also suffered from a lack of CAP:

> We were aware of the disagreements between Admiral Woodward, Commodore Clapp who was in charge of the landings, and with Northwood [RN headquarters], over the sea movement of troops around the Falklands. Even as bystanders, we were well aware of the dangers to the *Sir Tristram* and *Sir Galahad*, which it seemed our flagship was ignoring. There were plenty of pilots, who often were not flying, and plenty of aircraft.
>
> My opposite number . . . who'd been a fighter controller, phoned *Hermes* and told them that unless they launched CAP to protect *Galahad* and *Tristram* in the next ten minutes, it would be too late. He was told to mind his own business. But when finally CAP was launched, it was twenty minutes too late. . . . We were angry about that. We think there was an error in judgement [sic] in *Hermes*, which is a bit sad.[85]

The loss of the *Sir Galahad* was a high-profile disaster for Great Britain. Miscommunications and blunders that left the ship undefended resulted in 48 dead and 150 wounded.[86] The commander of the ground troops had refused to disembark his soldiers from the ship, despite being advised to do so by the BRN. By illustrating what happened to a ship that strayed outside the BRN's protection, this incident demonstrates the effectiveness of its scheme of defense for the beaches. Unfortunately, the great loss of life that resulted from this attack could have been avoided:

> I'd gone ashore to try to persuade 5 Brigade HQ to exercise some influence over the Welsh Guards officers who were refusing to leave the *Sir Galahad*. Their mistake was not getting off the ships. At the time I was extremely angry that they wouldn't take my word for it and obey my orders. There should have been air cover over Fitzroy that day, but HMS *Hermes* had steamed another hundred miles east to clean her boilers—against her captain's orders. You would have thought this was a bloody stupid time to decide to clean your boilers, but Admiral Woodward had ordered this, which meant that the CAP aircraft could only stay overhead for ten to fifteen minutes. Added to this, the CAP from *Hermes* were always stationed at high level, whereas *Invincible*'s CAP were always stationed at low level.

> Nobody has ever explained the reason for that, and it meant several Skyhawks got in under their defenses while *Herme*'s [sic] CAP was chasing other aircraft.[87]

The rest of the BRN ships that were lost in this battle were frigates and destroyers. This meant that the escorts did their job in protecting the landing ships, but the price was still significant. The BRN did not have a limitless supply of escorts. While potent, the Sea Dart and Sea Wolf combination had its limits, as demonstrated by the sinking of HMS *Coventry*. While guarding the mouth of Falkland Sound alongside the Sea Wolf–armed HMS *Broadsword*, four Argentinian A-4s attacked both ships in pairs. The first pair evaded *Coventry*'s Sea Dart by flying low and approaching from the land. When they came within range of *Broadsword*'s Sea Wolf, the system locked up and shut down; the two attacking aircraft had approached the ship at an equal distance from the weapon's sensor. Designed to engage the closest threat, the Sea Wolf system could not decide which one to engage and shut down with scant seconds to fix the problem. The one bomb struck Broadsword's flight deck, wrecking the embarked helicopter even though it failed to explode. The second pair of A-4s approached less than two minutes later; *Coventry*'s Sea Dart again could not track or engage them. *Broadsword*'s Sea Wolf was again operational, but was blocked from engaging by the hull of *Coventry*. Three bombs struck *Coventry*, two of which exploded in the engine room and just forward of the combat information center, respectively. The damage was catastrophic. Suffering 19 dead and a further 30 injured, the ship quickly took on water, and began listing until it capsized and sank 20 minutes later.[88] Theoretically, if *Coventry* had a Sea Wolf system on board, it could have employed the system in a way that could have avoided the problem of being blocked by its own superstructure. This issue arose in large part because the layered defense was cobbled together from multiple ships and close-in defense is best provided by each ship for itself, rather than by others.

The *Coventry* had been on station for days, successfully engaging many enemy aircraft that approached straight from the sea, but also making itself an important target for Argentina in the process.[89] While its position relatively close to Pebble Island made its acquisition from the air a tough prospect (Argentinian aircraft flying low over the island had only seconds to visually acquire the ship), the Argentinian pilots knew where it was located, and their knowledge negated that advantage. Furthermore, *Coventry*'s safety was a secondary concern for Woodward, who was preoccupied with the air threat, including from Exocet, to both the carriers and the landing ships:

> We were in the wrong place, ten miles north of Pebble Island. The Argentines knew exactly where we were; as they flew up Bomb Alley we were sitting ducks. Our skipper asked to be moved further north between West Falkland and Argentina, to take out aircraft before they arrived at the Falklands and give us a fighting chance. But the admiral refused. My friend was a Spanish radio intercept translator, and said they called us something like: "The British bastard north of Pebble that isn't letting anyone get through," a notoriety we could have done without. Our captain again asked to be moved north so our radar could shoot down aircraft coming at us from over the land. Admiral Woodward's signal said, "You are to remain in your current operating position. You are a missile target."[90]

The BRN layered defense addressed two of the three Argentinian threats: Exocets and low-flying aircraft. Rather than taking on the third with only defensive measures, the BRN took limited but decisive action.

Active Measures: Countering the Surface Threat

The third threat with which the BRN had to grapple was the Argentinian Navy's surface fleet. As noted above, it possessed the largest concentration of Exocets in the arsenal. The BRN never faced the threat of a large, concentrated Exocet attack, though.[91] This was in part because the BRN took action to negate this threat relatively soon after the start of hostilities. It demonstrated the vulnerability of the Argentinian fleet to submarines by sinking the high-profile ship *General Belgrano* with the nuclear attack submarine HMS *Conqueror*. After this attack, the most potent anti-shipping components of the Argentinian fleet, its carrier and destroyers, stayed in port. The Argentinian carrier and its other remaining surface elements lacked a proper frigate screen, which would have provided adequate protection against submarines.[92] The Argentinian Navy as a whole was particularly vulnerable to British SSNs, because there were few areas around the Falklands that SSNs could not reach.[93] After May 4, the main force of the Argentinian Navy, Task Groups 79.1 and 79.2, withdrew to home waters because of the *General Belgrano* incident and the fear of further attacks. Thereafter, the Argentinian fleet was limited to ASW, surveillance, defense of Argentinian ports and their approaches, and electronic warfare tasks. The Argentinian Navy's carrier aircraft were dispersed to land bases, from which they took part in the air campaign.[94] The Argentinian fleet became mainly a passive threat, a "fleet in being"—which the British, however, could not totally ignore.[95]

The SSN turned out to be a remarkably efficient counter to the Argentinian Navy's low-level air-attack tactics. The *General Belgrano*'s "loss seems to have blunted the entire Argentinian naval effort."[96] The only downside to this tactic was the apparent public relations backlash against Great Britain for the considerable loss of life inflicted outside of the British-declared Maritime Exclusion Zone. One must speculate how effective this would have been against a more motivated foe. The BRN, however, had several other SSNs in the area, which could have inflicted crippling damage on the Argentinian fleet.[97]

Argentina had, especially with its air force, demonstrated the ability to inflict significant harm on the BRN. They had inflicted "serious but not decisive losses on the fleet."[98] There were, however, only two instances in which their efforts were decisive. The sinking and damaging of the BRN's picket destroyers and frigates did not inhibit the BRN's achievement of its objectives. Argentina was only able to cause truly decisive damage to just two ships—the *Atlantic Conveyer* and the *Sir Galahad*—which had a negative effect on the BRN's land campaign. Despite the pounding that the BRN's ships received, it was the BRN that won the attrition game by shooting down so many aircraft. They had imposed a level of attrition that Argentina could not support; even though the BRN was suffering, the Argentinian Air Force and Navy could not maintain the tempo of operations.

The British Royal Navy's Organizational Ability

The reason for the BRN's reliance on numbers for its adaptation lies with how it was organized. The BRN's ability to mount a large, relatively well-supported operation is a clear indication of its strength in resources. On paper, the BRN task force was well organized to meet the threat. The task force commander's subordinates held distinct responsibilities (anti-air, anti-surface, antisubmarine warfare, etc.) and had assets assigned to them with which to carry out their duties. In practice, however, improperly centralized control of the task force interfered with the BRN's prosecution of the air war.

The BRN used numbers to counter the Argentinian innovation. While there were flaws in its execution—employment of SHAR being a primary example—this was the best that could be done when faced with a highly technical problem in a very short timeframe. Despite losing a significant number of ships, the operation succeeded because this adaptation played to the resource strengths of the BRN. Technology by itself was significant but not decisive. The lack of AEW and CIWS caused a lot of problems. No part of the BRN's operation was ideally suited to counter the Argentinian threat.

The British Royal Navy's Lack of a Dedicated Organizational Subunit or Top-Level Review Panel

Why did the BRN fall back on its resources rather than formulating a more efficient response to the problem? The answer lies in two different perspectives on the adaptation requirement. In one way, the requirement was competitive, with more than one subunit "managing" the air war simultaneously. In another, the requirement was shunned, because the subunit primarily responsible for countering the innovation was not properly equipped. In the first perspective, the BRN's organization clearly defined its subunits and their responsibilities, but was undermined from within. In the second perspective, the kinds of weapons used by the task force were able to defend only portions of the fleet completely at any one time. I will expand on each perspective in turn.

Looking at the BRN's task force organization, we see that officers subordinate to the commander each held responsibility for managing the surface, subsurface, and air defense of the fleet. Captain Black of the *Illustrious* was the officer designated for the air warfare job. Commander Ward was the senior SHAR pilot in the BRN, and was the most knowledgeable officer on that aircraft's employment. As Argentina's low-level mode of operation imposed its adaptation requirement on the BRN's organizational arrangement, it became clear that the operation was well-constructed in terms of subunits needed to meet the new threat. As the cases in this volume demonstrate, an organization that is able to get one subunit—a dedicated organizational subunit (DOS)—to focus on problem-solving is more likely to be successful. In this case, however, the actions of the task force commander largely undermined the BRN's efficient arrangement of subunits.

The SHAR contributed not as a method of stopping Argentinian air strikes, but as a way of increasing the attrition on Argentinian air forces. This is clear from the method of their use. While the SHAR was able to achieve great success, it was limited in its effectiveness by two things: the lack of AEW, and the mistakes made in its employment. The former is indisputable: no AEW aircraft were available until immediately after the conflict ended. Limited to their own and the picket ships' radars, the SHAR could not intercept Argentinian aircraft, especially the Exocet-armed Super Entendards, in time. Their sensors were simply not up to the task. The intelligence assets in place near the Argentinian coast could only accomplish so much; they could alert the fleet that an attack was on its way, but could not provide specific threat information such as the direction and speed of the incoming aircraft.

As to the employment of the SHAR, what the task force commander delegated he sometimes took away, most notably the SHAR available on his flagship *Hermes*. Sometimes these valuable aircraft were unavailable for defense of the fleet at crucial moments, or were positioned to intercept Argentinian aircraft only after they had made their attack runs. Furthermore, he positioned surface units, most notably the anti-air Type 42 destroyers, for early warning duty that might have been more effectively carried out by SHAR patrols. This demonstrates that even in militaries, division of responsibility is not always rigid, sometimes to the detriment of the organization. Admiral Woodward, a submariner, may have been better served in the air war if he had left it to those who specialized in it. As noted above, personnel in both the BRN and the British Army with knowledge of how the air warfare was conducted were often perplexed and frustrated by the use of the SHAR in providing air cover and the practice of trying to attrite the Argentinian Navy after it had already attacked the BRN's fleet. While the attrition strategy was effective in that Argentina was unable to keep up its initial tempo of attacks, this begs the question: if it was possible to accomplish attrition after the air strikes, why not do so before them, saving some ships and casualties in the process?

Woodward's exercise of authority over employment of the SHAR effectively put another subunit (the task force commander) within the adaptation requirement (see figure 5.4). This in part explains the inefficiency of the BRN's response to the Argentinian innovation.

The second perspective on the adaptation requirement is that the BRN was not equipped with the proper weapons to combat the Argentinian innovation. Given that naval combat had become a very technical business by 1982, this perspective is important because there was relatively little that the BRN could do to correct this arrangement in the time available. The Sea Wolf was the only weapon that was capable of engaging Exocets, and it did shoot down low-flying aircraft. Of the thirty-five warships in the BRN's operation, however, only three possessed the Sea Wolf, two of which were Type 22 frigates. These ships were a crucial piece of the puzzle in combating the Argentinian Navy's low-level air threat. Many of the BRN's other weapons were capable of engaging low-flying aircraft, but were either not very good at it or were limited due to guidance and detection shortcomings. The Type 42s, with their Sea Dart missiles, effectively shut down the medium- and high-altitude options for the Argentinian Navy, making attacks at anything but very low altitude

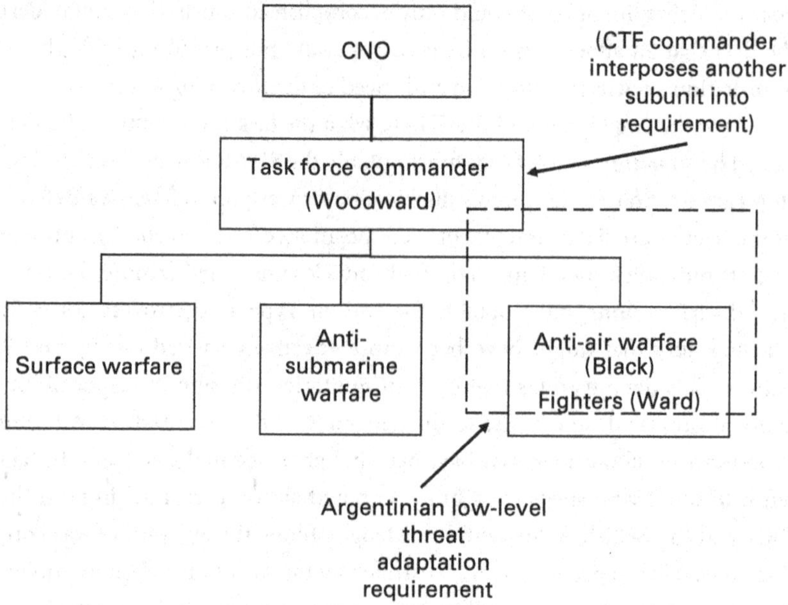

Figure 5.4
The British Royal Navy's organization relative to the adaptation requirement.

completely ineffective; however, its inability to engage low-level targets meant that the Type 42s needed a complementary system—in this case Sea Wolf, which was only deployed on another ship—for adequate protection.

When it was available, the partnership of a Type 42 ship and a Type 22 ship was deadly. Until it was sunk, the *Coventry* proved to be the bane of Argentinian pilots trying to thwart the landings at San Carlos Water. The Super Entendard, even with its air-launched Exocet, had to be very careful to avoid the BRN's Sea Dart system. It was precisely this effectiveness that made the *Coventry* a target for Argentinian aircraft, which eventually exposed the technical defect in *Broadsword*'s Sea Wolf, resulting in the former's loss. Furthermore, as the *Atlantic Conveyer* incident proved, chaff was effective in luring the Exocet away from its first acquired target; however, this did not destroy the missile, which had potentially disastrous results for other ships. The ship that launched the chaff protected itself, but potentially exposed others to the missile strike. Figure 5.5 represents the adaptation requirement relative to the weapons deployed by the BRN.

The BRN had plenty of anti-aircraft weapons, but not the right ones. This led to the shunned adaptation requirement, and partially explains why the BRN relied on numbers rather than countermeasures to address the

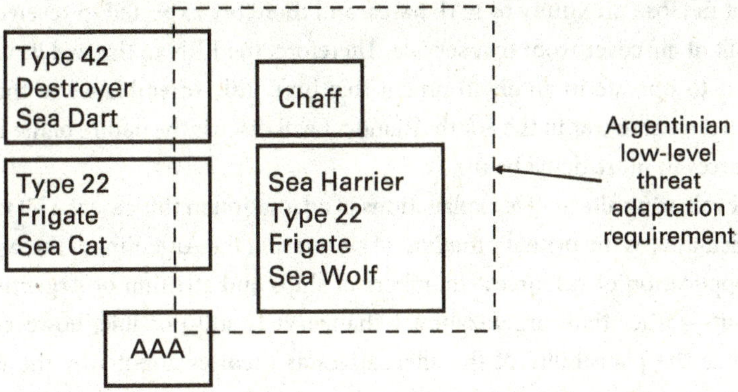

Figure 5.5
The arrangement of the British Royal Navy's weapons relative to the adaptation requirement.

Argentinian mode of operations. It should be noted that even though the Sea Wolf was the only BRN missile capable of engaging an Exocet missile, it never had the opportunity to do so, because the Argentinian Navy never got close enough to the carriers for the Type 22 frigates to fulfill their "goalkeeper" role. One can only speculate about why Argentina did not use the Super Entendard and Exocet combination to at least partially clear the way for conventional attacks on the BRN's ships in the San Carlos Water, especially the *Coventry*, as this was before the attack on the *Atlantic Conveyer*, when the entire inventory of air-launched Exocets was still available.

In this case, the BRN had a subunit designed to counter the threat that existed, but it was not properly equipped, which meant that in most cases it was ineffective. The BRN only had two of the ships most effective in countering Argentina's low-level threat—the Type 22 frigate with the Sea Wolf—so they could only be allocated to a couple of vital areas. This was not purely a resource issue, however. The BRN's operation was fairly large. The crux of the problem was that the resources were not properly oriented toward the threat that the BRN faced in the South Atlantic. In other words, the BRN went to war well equipped to face a medium-altitude Soviet-style threat with assistance from the RAF in both AEW and fighter interceptor tasks, but actually faced a low-altitude Argentinian threat, far from any British-controlled bases. Soviet anti-ship missiles were long-range and high-flying, requiring a missile like Sea Dart to counter them. Soviet bombers, which carried those anti-ship missiles, were best countered by long-range fighters, which the new British light carriers could not deploy. The BRN was increasingly expected to face the Soviet threat in close proximity to RAF bases, and therefore expected to receive the benefit of air cover from that service. Therefore, the BRN optimized its capabilities to operate in an environment that bore little resemblance to that in which it went to war in the South Atlantic. I will discuss the nature of the BRN resources in more detail below.

Evaluating alternative explanations of adaptation in this case is a bit more problematic, as the primary method of countering the Argentinian threat was the application of resources—numbers of ships and attrition of Argentinian aircraft—rather than organizational change. It is still possible, however, to evaluate the plausibility of the alternatives as theories about why the BRN simply relied on its resources rather than devising a more comprehensive approach to the problem.

As might be expected, civilian "intervention" had an effect on the course of the war, but not in a way that directly affected the adaptation process. The BRN came up with its own adaptation without civilian intervention. Civilian efforts did have a decisive effect in two important areas, though. First, before the outset of hostilities, their diplomatic efforts choked off Argentina's supply of French-made Exocet missiles. This greatly limited the extent of the air-launched ASM threat, though it did not eliminate it. Second, civilian leadership permitted the sinking of the Argentinian cruiser *General Belgrano* before it became a threat. This demonstration of force kept the rest of Argentina's surface fleet—along with its Exocets—in port. In any other case, however, this would not be a reliable method of adaptation. If Argentina had acquired a Soviet ASM instead of the French Exocet, for example, British diplomatic efforts to interrupt its supply would have been much less effective.

The competitive theory of adaptation success likewise holds little explanatory power in this case. The BRN was essentially on its own against both the Argentinian Navy and Air Force. Like the U.S. Air Force's Tactical Air Command in Chapter 4, the BRN was arguably the loser in a long budgetary competition with another organization, in this case the RAF. One of the BRN's motivations during the campaign, however, was to demonstrate its relevance as an expeditionary force and its ability to operate without RAF assistance. In this way, competition may have driven the BRN to successfully adapt. On the other hand, the goals of survival and victory were more pressing.

Resources

Resources were the BRN's primary source of strength during the Falklands War. This strength must be qualified to a great degree, however; while the

BRN had many ships, it was short on the weapons systems needed to counter Argentina's low-level threat. As discussed above, this was largely due to organizational factors, namely the decision to construct the BRN to face the Soviet threat to NATO's northern flank. Thus, while the BRN had a lot of warships, could call upon a substantial fleet of auxiliaries, and relied upon highly-trained and motivated personnel, it suffered a critical resource shortage in the air defense of the fleet.

The simple lack of these assets, such as AEW, CIWS, and more fixed-wing aircraft, does not tell the whole story, however. As noted above, the origin of the BRN's lack of these systems was largely a product of the retreat from the "East of Suez" strategy. In a tough budgetary environment, the British defense establishment and national leadership had made choices beginning after World War II that led the BRN to its organizational position immediately prior to the Falklands War. Great Britain had retreated from its global network of colonial possessions, which removed the justification for a superpower-class navy. The expeditionary East of Suez mission had enabled the BRN to keep its carriers and amphibious warfare capabilities, but in the face of economic austerity, the country had rejected this role as well by the late 1960s. The BRN was then unable to justify the retention of its carriers, and the last one was decommissioned in the 1970s. The dedicated naval AEW capability went with it. As the BRN shifted its focus toward the NATO mission, it came to rely on the RAF for air cover. Therefore, when the need for Operation Corporate arose, the BRN sailed without important capabilities, because they did not have the appropriate equipment for the task, and because there were no nearby bases from which the RAF could adequately provide air defense for the fleet.

The changing and narrowing of the BRN's core mission was ultimately responsible for its technological shortcomings in the Falklands War. While many ships and aircraft received minor improvements in an extremely short period of time, AEW and CIWS were too complex to develop on such short notice. Thus, many weaknesses in the BRN's capabilities could not be redressed in time to affect their operations.

The BRN's available resources went a long way toward addressing its shortcomings due to the lack of a DOS. Budget cuts and changes in the BRN's mission had begun to erode its strength. This process was not yet complete, however, and the BRN still retained the infrastructure and expertise of a Great Power navy. It was also able to recall some ships from the breakers and the auction block. Relative to its opponent, it had a large number of operational ships still in its possession. Additionally, the BRN had the facilities to maintain those ships, and to make the ones that were not immediately available quickly ready for sea. The organization was still well-placed to support the equipment and to send replacements to the war zone in time to impact the battle. Furthermore, the number of ships taken up from trade (STUFT) are indicative of the organization's ability to implement plans to support its operations.

The expertise of the BRN's personnel figured heavily in its resource advantage. While Argentinian aircraft would always outnumber the SHAR complement, the BRN's pilots, supplemented by carrier-qualified RAF pilots, were well trained, and their aircraft were superior in Air Combat Maneuvering (ACM, more commonly referred to as "dog-fighting"). In addition, the BRN retained the logistical personnel to accomplish the aforementioned modifications, maintenance, and repairs that kept its ships in service. Finally, the men of the BRN, while not combat veterans, were well trained even before the first shot was fired in the Falklands. Their training made the war seem much like "normal exercises."[99]

Conclusion

The BRN's ability to meet the Argentinian threat in the Falklands War was based upon a high level of resources in ships and personnel expertise. It was in a disadvantageous position, however, in terms of the adaptation requirement and the lack of a DOS. Thus, Argentina's innovative mode of operation was effective in sinking the BRN's ships. The primary reason why this innovation was not tactically or strategically decisive in the war was poor execution by the Argentinian forces. Rather than concentrating on the more important transport vessels and carriers, Argentinian pilots most often attacked the first target available. If they had waited a few more moments before pressing home their attacks, BRN losses could have been much more severe, perhaps decisive. This, as much as the British adaptations, can account for the poor Argentinian showing in the operational and strategic outcomes of the war. Few military leaders desire the success of their operations to depend on repeated and persistent mistakes by their adversary, though, for those are in large part outside of their realm of control.

This case suggests that as military methods become more technologically-focused and complex, adaptations become more time-consuming, and hence difficult to accomplish in a relevant time period. Technology optimized for one method of warfare may not be easily modified to adapt to another. The exemplar in this case is that the primary air defense for the BRN in this time period—Sea Dart and SHAR, with no AEW—had no way to engage low-altitude threats, and it was impossible to modify the missile system or cobble together AEW in the time available. Therefore, innovating in ways that maintain a basic level of flexibility is increasingly important compared to specialization in the era of high-technology warfare.

6

The Imperial Japanese Navy and the U.S. Submarine Menace in World War II

THE CASES IN this volume have thus far analyzed adaptations to innovation that have, to varying degrees, ultimately been successful. In contrast, the case of the Imperial Japanese Navy (IJN) and its struggle against the U.S. Navy (USN) and its "Silent Service" serves as an example of an almost complete failure to successfully adapt. The U.S. submarine offensive started off slowly and achieved limited results in its first two years. With gradual improvements, it managed to cut Japan's lines of communication during 1944. The IJN's adaptations were so ineffectual, and its anti-submarine warfare (ASW) forces were so poorly organized and resourced, that its efforts to protect Japan's sea-lanes of communication (SLOC) foundered even as U.S. submarines were largely ineffective. As the USN increased its skill and improved its weaponry, the IJN failed to adapt while there was still time to do so.

It is impossible and undesirable to deny that Japan chose to launch a war against an adversary that had an insurmountable lead in resources of almost every kind. As in most cases, however, analysis based solely upon resource disparity only tells part of the story. It is important to note, for example, that the percentage of effort that the United States committed to its submarine offensive relative to the rest of the USN's strength was rather modest. Only 1.6 percent of USN wartime strength (16,000 men) accounted for 55 percent of Japanese shipping losses (6.1 million tons displacement total).[1] Unlike the German U-boat offensive in the Atlantic, during which the *Kriegsmarine* committed a maximum of 100 boats operational at any one time, the USN only had 50 submarines operational in the Pacific at any one time. This included approximately 22 in transit to and from their assigned patrol zones.[2]

Japan was particularly vulnerable to commerce warfare. While its 1941–1942 offensive had secured the raw materials it needed, these goods were spread out over thousands of islands up to 3,000 miles distant from the core of the empire. Japan was dependent upon its merchant marine to bring those resources home and support the garrisons that protected them. Prior to the war, however, the IJN paid little attention to commerce protection, regarding it as too "defensive" in nature, especially given that its overall war planning relied upon a short-war strategy. It also failed to anticipate how the USN would use its submarines: "Japanese submarines were trained to sink warships; it was expected that foreign boats would do the same."[3]

These factors show that the disparity in resources alone fails to explain the outcome of the USN's commerce war against Japan. Resources were important, but overwhelming numbers of USN submarines certainly did not smother the IJN. Like any organization adapting to a wartime situation, the IJN faced critical choices early in the conflict. The poor choices it made and the effects of its pre-war policies had a decisive impact on its ability to counter the U.S. submarine threat, above and beyond its lack of resources.

In order to understand why the commerce war in the Pacific proceeded as it did, it is necessary to look at interwar defense policy, this time in the area of IJN shipbuilding. The IJN, which focused on a short-war strategy, neglected missions such as commerce protection and ASW, as reflected in its shipbuilding choices. I will then examine the U.S. innovation. It was indeed a wartime innovation, as USN submarine doctrine did not envision a war on commerce, but rather as auxiliaries to the battle fleet. Hours after the Pearl Harbor attack, the USN changed its policy on commerce war, but almost two years passed before the IJN faced "wolf packs" of submarines with effective torpedoes and aggressive skippers. After the nature and effect of the USN innovation are explored, IJN measures to adapt to the submarine menace will also be examined. While Japan deployed the trappings of an effective response, such as convoys, ASW frigates and aircraft, and sonar and other electronic detection equipment, they did so in an ineffective and uncoordinated manner. The IJN's ability to organize to meet the threat was poor. When it did attempt to reorganize, there were not sufficient resources for the task. Not surprisingly, this led to the fairly quick destruction of the Japanese merchant marine once the U.S. offensive gained momentum.

Sowing the Seeds of Disaster: The Imperial Japanese Navy in the Interwar Years

As a resource-poor island nation with imperial ambitions, Japan spent much of the late nineteenth and early twentieth centuries in a quest for autarky. Japanese leadership realized that the country relied too much on outside powers, especially the United States, for its raw materials. This vulnerability constricted Japan's freedom of action in foreign affairs; its ambition would soon collide with those of rival powers, and those powers could use Japan's vulnerabilities to interrupt its expansion.[4] This basic problem drove Japanese war planning, but that planning did not foresee defending Japan's autarky once it had been gained.

In the interwar years, Japan based its defense policy on two factors. First, its most likely adversary would be the United States. Second, Japan could not compete with the United States in terms of resources and industrial output. Therefore, any war with the United States would have to be a short one. In its quest for autarky, Japan pursued a defense policy that relied upon a short war and a decisive sea battle. The empire's commerce protection measures were indicative of this reliance. The entire point of going to war with the Allies was to gain resource independence, but the protection of those resources while in transit from the outer empire to the Home Islands was completely neglected in the planning during interwar years and the first year of the Pacific War itself. The IJN's shipbuilding policy and its administrative structure had a significant negative impact on the Japanese wartime adaptations.

Japan's yardstick for measuring its naval strength was the United States, because it seemed to be the most likely and strongest future opponent. In the era of the Washington Naval Treaty (1922–1932), the IJN out-built the USN twice over.[5] In anticipation of the eventual collapse of the treaty restrictions, however, the United States also expanded its shipbuilding in 1934. In their naval expansion programs, both countries produced new capital ships, but the U.S. program concentrated greater effort on auxiliary and supporting ships. The USN thus entered the renewed naval arms race with a much more balanced fleet than the IJN.[6]

The IJN building program reflected Japan's short-war strategy. Since Japan's war planners realized that it could not compete with the industrial potential of the United States in a long war, they concentrated efforts on achieving qualitative superiority in combat units. The IJN thus emphasized the super-battleship program and naval aviation. The result was a very effective but fragile naval air arm, and two battleships that outclassed everything else afloat.[7] As David Evans and Mark Peattie note, "When the war finally came . . . the ships that these building programs provided would be too few and too many of the wrong kind."[8]

To illustrate the IJN's lack of emphasis on building ships other than battleships and carriers, it is helpful to do a numerical comparison between the U.S. and Japanese building programs. The commerce war in World War I had demonstrated that the most effective ASW ship was the destroyer, based on its speed, armament, and flexibility. These characteristics also meant that the destroyer was in high demand; adequate numbers were needed both to protect the battle fleet and to accomplish other missions, such as commerce protection. Therefore, a useful metric for determining whether or not an adequate number of these vessels have been built would be the ratio between numbers of destroyers and capital ships. The higher the ratio, the more excess capacity a navy has for protection of vessels outside of the battle fleet. During the years of 1937–1941, the United States built 6 capital ships (2 battleships and 4 carriers), and 68 destroyers, for a ratio of 11.3 destroyers for each capital ship. During the same period, Japan built 12 capital ships (one super-battleship and 11 carriers, with another super-battleship completed the next year), and 37 destroyers, for a ratio of 3.1 destroyers for each capital ship.[9] The IJN had barely enough destroyers to protect its battle fleet, let alone its commerce.

The policy of achieving qualitative superiority cost the IJN dearly in perhaps its most important mission areas. The best example of this cost is the coastal defense vessel, a group of similarly capable ASW frigate classes called *kaibokan*. These vessels turned out to be the best ASW platform the IJN had during the war. Designed as 1,200-ton vessels equipped with sonar and large numbers of depth charges, they were reduced to 860 tons to defray the cost in materials of the battleships *Yamato* and *Musashi*. Only four were built prior to the war.[10]

The most relevant criticism of the IJN's super-battleship strategy does not concern the building of the super-battleships themselves, but the consequent diversion of the Navy's attention and the nation's resources from one of the most critical strategic problems that the IJN faced: its utter dependence on its overseas sea routes and the need to acquire the ships, form the organization, shape the doctrine, and develop the training that would be most effective in protecting those sea routes.[11]

One reason the IJN was able to turn a blind eye to the commerce protection problem was its failure to anticipate that a war on its commerce was possible. While many U.S. submarine officers advocated the submarine as ideal for commerce war, U.S. doctrine envisioned them primarily as auxiliaries to the fleet. In other words, U.S. submarines would scout ahead of the main battle fleet, hunt capital ships, and otherwise support the aim of bringing about a decisive sea battle.[12] IJN doctrine called for using its submarines in the same way.[13] The IJN was guilty of a sort of mirror-imaging between its submarine force and that of the United States: "Seeing that force as directed primarily against U.S. fleet units, the Japanese naval high command apparently assumed that their opposites in the U.S. Navy harbored similar plans, as indeed they did, up until the beginning of the war."[14]

The neglect of commerce protection was not unique to Japan. The United States and Great Britain were both guilty of the same indifference. The USN and the British Royal Navy (BRN), however, both had institutional memories of the requirements of convoy protection gained from their experience in World War I. While the IJN contributed destroyers to commerce protection in that war, and those ships performed well, the experience left little impression on the service. The most significant event of that war from the IJN's perspective was the Battle of Jutland: "Japanese understanding of the rigors of commerce protection against submarines was limited in time, distance, and practical experience."[15]

The IJN acknowledged the theoretical problem of commerce protection, but took no practical action to address it. In addition, it seemed to ignore some obvious warning signs. For example, an IJN submarine penetrated home waters during an October 1940 exercise, "sinking" 133 merchantmen. These were the same seas in which the USN would devastate Japanese shipping four years later. The primary lesson drawn from this incident, however, was that submarines were vulnerable to radio direction-finding equipment.[16]

Besides the lack of priority given to ASW in IJN shipbuilding, organizational culture and structure both conspired against an effective policy. Most IJN officers had disdain for the commerce protection mission. They had no experience in, and little understanding of, convoys or ASW. Professionally, it seemed to be a "marginal and unrewarding trade."[17] Furthermore, "the defensive character of both convoy escort and ASW, and the tiresome, repetitive duty that they entailed, gave them short shrift in a navy whose traditions had held that offense was the best defense."[18] Like the BRN at the outset of World War I, the IJN was prejudiced against overly "defensive" missions, a bias it would have to overcome if its commerce was threatened in a long war.

The organizational structure of the IJN reflected the general feeling toward ASW and commerce protection within its ranks. Generally, the IJN viewed commerce protection as a coastal defense issue. Geographically, this limited the mission to areas north of Taiwan and in the northeast Asia littoral. Forces detailed to this mission would come from rear-area defensive units and wartime construction. Fleet units would only be detached from front-line service if necessary. Also, "the 1941 operational plan did not make special reference to commerce protection."[19]

In addition, no organization within the high command existed that had exclusive responsibility for commerce protection. Until 1941, there was only one officer on the planning staff for "rear area" defense. On the eve of war, this was increased to two officers. Also, "there was no major line command for commerce protection, and none was established until the war was four months old."[20]

The Japanese pre-war naval policy reflected the IJN's neglect of commerce protection in almost every conceivable area: "Thus, on the eve of the Pacific War, the Japanese Navy had given almost no thought to the possibility of a submarine campaign in its rear areas, and commerce protection . . . was a mission to be improvised as the need arose."[21] Japan would pay dearly for an oversight on such a grand scale.

The Innovation: The U.S. Navy and Submarine Warfare in the Pacific

The U.S. submarine offensive against Japan started less than twelve hours after the attack on Pearl Harbor on December 7, 1941.[22] The years 1942 and 1943, however, turned out to be a period of learning and adjustment for the USN. The U.S. submarine war did not start in earnest until 1944.[23] That was the year that USN adjustments to its submarine campaign became decisive.

In 1942, the U.S. submarine force was split between several tasks: raiding Japanese commerce, targeting Japanese warships, and opposing Japanese amphibious operations. Besides this lack of focus, the force itself suffered from two major weaknesses: "Much of the submarine failure in 1942 could be laid to poor skipper performance and poor torpedo performance."[24] During the interwar years, submarine captains were promoted and recognized for their ability to remain undetected, not their aggressiveness. The peacetime USN valued attention to detail in paperwork, discipline, and caution. Officers with these traits were promoted to command, while more aggressive officers were punished for taking risks in exercises. In the first full year of the Pacific War, this policy yielded many unproductive patrols. Skippers were persistently timid in the handling of their boats. Thirty percent of Pacific Fleet skippers lost their commands during 1942. They were relieved primarily because of unproductivity—i.e., not attacking or sinking enough enemy vessels—but also due to poor health and battle fatigue.[25]

In addition, the primary U.S. anti-shipping torpedo, the U.S. Mark XIV, was ineffective. The depth at which it traveled from the submarine to the target was variable and set before firing, but the torpedo took a deeper path than the setting. Submarine captains quickly recognized that something was wrong when torpedoes missed targets that were extremely close and had sufficient draft so that the weapons should not have passed underneath them.

Test-firing against nets in harbor proved that the weapons were running too deep. Directions for corrective action—setting the running depth more shallow—were soon promulgated throughout the submarine force.

There was a more insidious problem with the Mark XIV, however, that went unaddressed throughout 1942. A large proportion of these weapons failed to detonate when they struck targets. The Mark XIV had two kinds of exploders: contact and magnetic. The former operated by detonating the warhead when it struck the target's hull, while the latter was designed to detonate when it sensed the disturbance in the earth's magnetic field generated by the hull. The contact exploder failed frequently, and the magnetic exploder did not work at all. There was "almost universal belief" among the submarine force that the exploders were faulty, but no live-warhead tests were conducted.[26]

At the start of the Pacific War, Pacific Fleet submarines spent much of their time hunting Japanese warships, which detracted from the task of disrupting commerce. Once Japan overran the Philippines, U.S. boats operated from Pearl Harbor and from two bases in Australia, Freemantle and Brisbane. Patrols originating in Australia tended to chase capital ships, while those from Pearl Harbor went to Japanese waters on commerce raiding missions. Even though the latter consisted of only 15 percent of the war patrols that year, they accounted for 45 percent of all ships sunk.[27] Losses were comparatively light at seven boats, but four of those were due to grounding in shallow waters; only three perished due to IJN countermeasures.[28]

Overall, the Pacific Fleet leadership, both submariners and higher-level commanders, did not take to heart the lessons of the Atlantic U-boat campaign against Allied shipping. Instead, they divided the submarines' attention among multiple tasks: "The U.S. submarine force was divided and shunted willy-nilly on missions for which it was not suited, while the bulk of Japanese shipping sailed unmolested in Empire waters and through the bottleneck in the Luzon Strait [in the Philippines]."[29]

It is not surprising, given these weaknesses, that Japanese shipping losses in 1942 were not particularly worrisome. During 350 war patrols from Pearl Harbor, Freemantle, and Brisbane, U.S. submarines accounted for 180 sunken ships totaling 725,000 tons displacement. German U-boats accounted for a similar amount of tonnage in just February and March of 1942.[30] Therefore, while the USN accomplished more than the *Kriegsmarine* in its total war effort, in 1942, it still had a lot of work to do in order to catch up to the level of Germany's efficiency and ability to sink ships.

The U.S. submarine offensive gained momentum in 1943, although it did not yet reach its full potential. In every aspect, there was a marked improvement over operations in 1942. More boats patrolled Japanese waters—50 percent compared with just 15 percent of the force in 1942—reflecting a greater concentration on merchant traffic over the hunt for enemy capital ships. Pacific Fleet submarines also experimented with wolf-pack tactics: rather than hunt targets singly, a group of submarines would coordinate its attack on a target (e.g., a convoy) to overwhelm the ability of any escorts to effectively defend their charges. This coordination of effort meant that submarine skippers had to sacrifice their much-prized autonomy. Most captains, however, were reluctant to adopt tactics that would limit their freedom of action. Furthermore, they were worried about friendly fire when submarines operated in close proximity to one another. These factors inhibited full implementation of wolf-pack tactics until 1944.[31]

The increased USN focus on Japanese commerce had a significant impact on shipping to and from the Home Islands. Compared to 1942, in which Japan imported 20 million tons of bulk commodities (e.g., iron ore), this figure fell to 16.4 million tons in 1943. Merchant losses totaled 335 ships for 1.5 million tons displacement, approximately double the total of the previous year. While Japan was able to increase its oil tanker tonnage, its shipbuilding for other types of transport was unable to keep pace with losses. The number of war patrols conducted by U.S. submarines was virtually identical, approximately 350.[32] This indicates that the U.S. submarine offensive was becoming more effective, rather than just increasing the number of boats on patrol. Increasing numbers of U.S. submarines were equipped with radar, giving them enhanced detection capabilities when surfaced. Also, the U.S. fleet made better use of available intelligence on Japanese shipping.[33]

In 1943, there were still downsides from the U.S. perspective. Increased contact with IJN countermeasures meant that fifteen boats were lost that year. There were still too many unproductive skippers in the submarine force, but reliefs decreased from 1942 levels because there were not enough trained officers to replace them. The effort to find and sink capital ships was largely wasted. Only one small carrier was sunk as the result of about 60 submarine–capital ship contacts generated by the search missions. Finally, problems with the Mark XIV torpedo exploders continued through much of the year. In September 1943, the submarine force declared the Mark XIV fixed, having formally deactivated and phased out the magnetic exploder and addressed the faults in the contact exploder.[34]

Thus, 1943 represented another "trial and error" period for the U.S. submarine force. It learned how to fight and received the equipment—working torpedoes, intelligence, radar, etc.—to do it. Therefore, as Clay Blair states, "In one sense it could be said that the U.S. submarine war against Japan did not truly begin until the opening days of 1944."[35]

The U.S. offensive in 1944 turned out to be "devastatingly effective" in terms of its effect on Japanese shipping. The U.S. submarine force sank 603 ships for 2.7 million tons, more shipping than the previous two years combined. Japanese bulk imports fell from 16.4 to only 10 million tons, with disastrous results for its industry and population. In part, the change in tactics increased U.S. effectiveness; when the IJN finally instituted convoys, the USN moved beyond experimentation and employed wolf-pack tactics in earnest.[36] The U.S. submarine force also redirected substantial effort to hunting down tankers, with great success. Japan built 624,000 tons of new shipping in 1944, but only netted a gain of 6,000 tons after accounting for its losses inflicted by U.S. actions. As a result, the flow of oil from the southern regions of the empire decreased to a trickle. The fuel situation became so bad that "Japanese leaders launched experiments in making oil from potatoes."[37]

U.S. submarine losses increased again from 1943 to 1944. In 1944, the USN lost 19 boats on war patrols. Even allowing for these losses and the retirement of 23 older or battle-damaged boats, however, the submarine force increased by 33. Also, with the Allied march across the central and southern Pacific, the additional new bases progressively closer to the core of the Japanese empire (for example, at Majuro and Kwajalein in the Marshall Islands, which shortened the distance to patrol areas by 2,000 miles compared to Midway Island) meant that patrols could be longer.[38]

The leadership and torpedo problems never disappeared entirely. Because there was always a shortage of trained officers to take command, captains with unproductive records had to be retained, especially with the amount of newly constructed boats entering the fleet. Torpedoes, once in short supply (as well as being defective), were now plentiful. Even though they were more reliable, their relatively small warheads only damaged many targets that otherwise would have sunk: "Had the torpedo warheads been larger—as large as Japanese submarine torpedo warheads—many more ships would have gone to the bottom."[39]

The U.S. submarine campaign tapered off dramatically in 1945, not because of a decrease in its effectiveness, but rather because merchant ships had become extremely scarce.[40] Submarines based in the Philippines "finally ran out of targets. During the waning days of the war, they roved the sea on the surface like destroyers, blasting away with deck guns at trawlers, sampans, and other small craft, most of them manned by natives, not Japanese. When the boats ran out of ammunition, they boarded the craft and blew them up with grenades and demolition charges."[41]

In 1945, there was one place left that had been virtually untouched by the submarine offensive. This was the Sea of Japan, which was protected by dense minefields at its northern and southern entrances. With the help of new sonar designed to detect mines, a large U.S. wolf pack penetrated these waters in June 1945. The nine boats of the wolf pack made it through the field without loss, and wreaked havoc among Japanese ships that sailed without escorts, zigzagging, or blackouts. These nine boats accounted for the destruction of 28 ships and 54,784 tons, and lost only one submarine. Again with the help of the new sonar, the remaining submarines successfully exited the Sea of Japan through the minefield to safety.[42]

The U.S. submarine offensive virtually destroyed Japanese merchant shipping. In 1941, the Japanese merchant marine retained personnel strength of 122,000. By the end of the war, it had suffered 116,000 casualties—27,000 killed and 89,000 wounded. The submarine force inflicted 69,600 of those casualties.[43] In terms of shipping tonnage, the merchant marine started the war with 6.4 million tons, and ended it with just 1.5 million tons, the majority sunk by submarines. These losses not only crippled Japan's ability to persist in the war, but also degraded the IJN's ability to fight because of the disappearance of its logistical support.[44] Even with these impressive totals, the claims made by U.S. submarine captains were on the whole wildly exaggerated. This brief narrative uses the numbers confirmed after the war. Admiral Charles Lockwood, who commanded U.S. submarines in the Pacific, chalked this up in part to the torpedo problem: "A total of 14,748 torpedoes had been fired. Had all these run, hit, and detonated as designed, the claims might well have been closer to the actuality [Admiral Lockwood] maintained. Few could disagree. The torpedo scandal of the U.S. submarine force in World War II was one of the worst in the history of any kind of warfare."[45] In some ways, Japan was lucky that its opponent continued to use faulty weapons and tactics through much of the campaign. It gave the IJN an opportunity to address its severe weakness in commerce protection.

Japan's difficulties did not arise from U.S. technological superiority either. If anything, the state of U.S. submarine technology made the submarine problem in the Pacific easier than that of the Battle of the Atlantic. U.S. submarines were capable, but not the most technologically advanced in terms of propulsion, sonar, or optics, and began the war severely deficient in armament.[46] The

IJN was not able to take advantage of this technological inferiority to successfully adapt. The next section explains this failure.

The Imperial Japanese Navy's Adaptation: Too Little, Too Late

For those committed to its execution, the U.S. submarine offensive was a hazardous endeavor. During the war, 52 submarines were lost, nine of those due to causes other than enemy action. This included two confirmed cases in which the submarine was sunk by its own malfunctioning torpedoes.[47] The submarine service suffered the highest loss rate in the USN. It lost 22 percent of its officers and men and 18 percent of its submarines.[48] The IJN, however, inflicted these losses with resources, organization, and tactics that were subpar at best. The USN's postwar assessment of its submarine losses best summarizes the Japanese ASW effort:

> The Japanese high command had given consideration to almost every phase of advanced anti-submarine warfare. Much thought and research effort was directed towards [ASW]. . . . Nevertheless, at the end of the war Japanese anti-submarine warfare was still waged without appreciable benefit from the results of their research effort. Lack of production facilities, inability to maintain adequate supply lines, and the necessarily large allocation of production of aircraft for the final defense of the Empire home islands greatly affected the anti-submarine forces, but there also existed a failure on the part of operating personnel to use what they did have to best advantage.[49]

Another post-war USN assessment stated that:

> It would do very well . . . for all submariners to humbly ponder the fact that Japanese anti-submarine defenses were not of the best. If our submarines had been confronted with Allied anti-submarine measures, the casualty list . . . would have been much larger and the accomplishment of Allied submarines much less impressive.[50]

The primary Japanese ASW ship was the coastal defense vessel, or *kaibokan*, known as a frigate to the USN. There were roughly four classes of these vessels, but their capabilities were virtually the same. These vessels had two 4.7-inch guns, and eleven 25-millimeter machine guns. Their most important armaments, though, were their twelve depth charge throwers and capacity of about 300 depth charges. The ships mounted "efficient" sonar. Top speed was 20 knots. The USN's post-war assessment of the *kaibokans* was that: "Altogether they were efficient anti submarine vessels, except that . . . their radar was inferior and they lacked the modern weapons for attack on submarines."[51] As noted above, these valuable ships were given low priority in the years prior to war. When the submarine problem became very serious, however, Japan began a crash-building program, completing 150 *kaibokan* before the end of the war.[52] While still dwarfed by U.S. production numbers,[53] this program indicates that there were resources available to commit to commerce protection. In 1942, the IJN commissioned the second super-battleship, the *Musashi*, and in 1945 commissioned the *Shinano*, a new carrier converted from the third super-battleship hull (the *Shinano* was sunk on its maiden voyage by a U.S. submarine). One can only speculate about how the resources that the IJN committed to these large vessels might have been used for the protection of Japan's shipping.

IJN fleet destroyers, in comparison with the *kaibokan*, were faster, better protected, and had better gun armament. They also mounted both active and passive sonar. They only carried about a tenth of the depth charges as the *kaibokan*, however.[54] Their capacity was approximately thirty, with twelve ready to drop from the stern and the rest stowed below. They did not have throwers like the frigates. Training was deficient. In ASW operations, the torpedo officer was solely responsible for employing the ship, independent of the commanding officer's authority. Yet, one captured torpedo officer revealed during interrogation that the last ASW training he had received was several years prior to the start of the war. His U.S. interrogators were incredulous that training was so lacking and that the captain of a vessel would cede so much authority to a subordinate, but determined that the POW was a credible source. These vessels played much less of a role in the ASW war than the frigates, first because the Combined Fleet jealously husbanded them for protection of capital ships, then because of attrition later in the war. In 1944, "due to the severe destroyer losses in the Japanese Navy, the function of anti-submarine warfare was relegated entirely to smaller vessels and. . . . Fleet destroyers were not expected to conduct ASW as one of their main functions."[55]

The IJN employed smaller vessels for ASW duties close to port. Patrol boats sailed the waters off important bases, channels, and other strategic locations. Sometimes they provided short-range escorts for large ships entering or leaving ports. Mostly they maintained a patrol at a particular station. These vessels were generally very small, displacing only up to 500 tons. The largest had passive sonar, radios, a small deck gun, and depth charge throwers. Most were less capable, with the smallest being sampans just large enough to carry a depth charge that could be dropped over the side. If they had sonar, it was a crude hydrophone lowered over the side. U.S. submarines still had to account for these vessels for a number of reasons. Their presence meant that remaining submerged was essential. Even though they were easily outgunned by the submarines, a surface attack on one would immediately draw the attention of other local forces. Their depth charges were potentially just as lethal as those carried by larger IJN vessels, but the vessels themselves were too small to engage with torpedoes: "Their slow speed and the weakness of their sonar equipment were what reduced their effectiveness."[56]

The IJN made heavy use of two types of sonar—hydrophones (passive listening) and "asdic" (active echo-ranging using sound waves). Its passive sonar was "fairly good," with the capability to detect U.S. submarines "at ranges of 2000 meters or more when the listening ship was stopped or proceeding at very slow speed."[57] Japanese active sonar was "mediocre," as its level of technology roughly corresponded to that of U.S. models produced in 1937. The temperature of the water at various depths affected the active sonar's transmission of sound significantly. This is due to the density of the water; deeper, colder water is denser than more shallow, warmer water. Sound waves typically do not transmit well across different water density layers. This allows submarines to hide from active sonar by positioning themselves below the boundaries between density layers. While Japanese scientists were fully aware of this phenomenon, IJN ships were not equipped with bathythermographs, which were devices used to measure water temperature at various depths, and "no tactical use was apparently made of the small amount of information of operational value issued by the Japanese Hydrographic Office."[58] Despite this low level of technology, IJN escorts manned their passive sonar continually and operated their active sets at least a quarter of every hour. Large portions of their already cramped ships were given over to sonar equipment.[59]

As noted above, surface search radar was an important technological aid to U.S. submarines in hunting their targets. It made shadowing a convoy and setting up an ambush easier and safer. On the defensive side of the equation, the IJN failed to take full advantage of radar to detect surfaced submarines. The IJN recognized the importance of radar, because it realized that U.S. submarines cruised on the surface frequently, especially at night; however, lack of equipment meant that many ASW escorts were never fitted with surface search radar. On the vessels that were equipped with radar, they were "in most cases unable to obtain satisfactory results, due to defective equipment and poor training of maintenance and operating personnel."[60] This allowed USN submarines to surface at night, and often in the daytime, with impunity. Being able to run on the surface allowed them to take advantage of their higher surface speeds, enabling them to cruise around their targets to execute multiple attacks from different directions over time.[61] Most IJN escorts had radar by December 1944, although by this time it was too late to make much of a difference.[62]

The IJN also made use of radar signal-receiving gear. This was used to detect radar signals from U.S. submarines, although it was not well developed.[63] The device noted the presence of radar signals, but did not give range or bearing information. As such, it served primarily as a warning device.[64]

The IJN's weapon of choice against the submarine was the depth charge: "Depth charges were placed aboard practically every type of ship capable of carrying them and even the smallest of patrol craft could generally be expected to have a few."[65] Fleet destroyers typically carried 30 of these weapons, while the *kaibokan* had a capacity of up to 300.

The depth charges themselves were copies of obsolete British designs. There were only two types in use during the war, the Type 95 and the newer Type 2. The former contained between 220 and 325 pounds of explosive, depending on the formula. The latter had up to 357 pounds. Both types were detonated by a "pistol"—a fuse activated by hydrostatic pressure, which was also of British design. As the depth charge descended through the water, an inlet allowed water to fill a cylinder. When the cylinder was full, the pistol activated and detonated the charge. Changing the size of the water inlet varied the depth at which the charge detonated. The Type 95 had only three depths at which it could be set: 98 feet, 98 feet with parachute, and 197 feet. The parachute was used when a vessel was traveling at slow speed, so that it had time to put enough distance between it and the explosion of the depth charge. The Type 2 had a greater choice of depth settings: 98 feet, 197 feet, 292 feet, 390 feet, and 480 feet. It was, however, "apparently not possible with either type of pistol to choose depth settings other than those enumerated above even though detonation at some intermediate depth might be considered desirable."[66] If the U.S. submarine was not at or near one of those pre-set depths, the charge was likely to be completely ineffective.

Late in the war, a third type of pistol, the Type 3, could be set to detonate at additional depth settings of 524 and 656 feet, but "issue to ships had not become general by the end of the war and no specific doctrine for its use had been developed."[67] Furthermore, and more significantly, because "the Japanese were unable to estimate the depth of a submerged target, no specific technique was developed for attacking submarines in deep submergence."[68] This lack of depth-finding capability represented a serious shortcoming in the IJN's ASW, as depth charges had to get fairly close to their targets in order to cause significant damage.

In comparison to Allied ASW weapons, the most glaring deficiency of the IJN was its inability to develop something similar to the "Hedgehog." This weapon consisted of a number of smaller depth charges launched simultaneously in set patterns. Their large numbers and contact fuses offset the drawback of their warheads' small individual size. Used heavily in the Atlantic against the German U-boat by the Allies, it could be launched ahead of the ASW vessel, as opposed to being thrown off the stern or over the side singly as was the case with more conventional weapons. The contact fuse had important implications for ASW tactics, most notably because the charges did not detonate unless contact with a submarine was made; therefore, the missed shots would not explode and interfere with sonar operation. A USN study estimated that a submarine was four times as likely to survive a depth charge attack as it was to survive an attack from a Hedgehog.[69] Even though their weapons were not state of the art, their use could have been much more effective:

> The Japanese failed to achieve optimum use of their principal anti-submarine weapon, the depth charge, simply by repeatedly setting their pistols too shallow. There are many instances reported . . . in which depth charge patterns detonated directly overhead without causing appreciable damage but where an increase in depth setting would probably have resulted in fatal or serious damage. Above all, the Japanese generally lacked persistence in both their search efforts and exploitation of a positive contact, once made. They were prone to accept the slightest evidence that a submarine had been destroyed and then depart. Many a U.S. submarine owes its escape to the Japanese predisposition to take the most optimistic view.[70]

In other words, with poor training and tactics, the IJN made inadequate use of the weapons it had available.

Minefields were an important part of Japanese commerce protection strategy. Like Great Britain in World War I, Japan laid extensive minefields in order to block likely submarine approach routes. The theory behind their use was that submarines would not be able to penetrate certain areas, enabling merchants along those routes to proceed without escort. The typical minefield was laid with a density sufficient to present a "ten percent threat," meaning that one in ten vessels passing through the field would be sunk. In practice, the effectiveness of the minefields did not live up to this expectation. Over time, the mines became less effective for three reasons: they became fouled (with, for example, sea life); their moorings snapped; or they sunk due to leakage, and the density of the minefield decreased. Furthermore, U.S. submarines frequently received intelligence on the location of minefields from captured enemy notices to mariners: "The most important of the Japanese minefields were therefore avoided as scrupulously by our submarines as they were by Japanese merchant traffic."[71] The introduction of FM sonar on U.S. submarines further degraded the effectiveness of minefields, as the description of the Sea of Japan wolf pack above describes.[72]

The IJN made extensive use of aircraft for ASW; however, a number of factors limited their effectiveness. Equipment quality and personnel training were the most significant problems. For example, many Japanese ASW aircraft were equipped with radar, but pilots were not properly trained or confident in its use. In fact, the IJN viewed radar, which might have detected many U.S. submarines that routinely operated on the surface, as a backup for visual detection. The latter was still considered more reliable except in times of low visibility or at night: "Towards the end of the war, Japanese radar-equipped planes were making numerous contacts with our submarines but few attacks resulted."[73] This indicates that the IJN was not taking advantage of the radar capabilities that it did have.

The IJN used aircraft as both direct cover and for sweeping ahead of convoys.[74] When contact with a submarine was made, ASW aircraft would immediately attack. The pilot would drop one or two depth bombs on the target, mark its estimated position, and then report the contact. Sometimes he might visually guide a surface vessel to the target. As long as fuel remained, the aircraft would stay on station: "Although many contacts in the early days of the war were broken off after a few hours, the accepted doctrine in the latter part of the war was to continue following the contact for two or three days."[75]

IJN aircraft did make heavy use of the Magnetic Anomaly Detector (MAD). The IJN developed this device in late 1943, and it entered service in March 1944 in both army and navy aircraft. The MAD is used to indicate the presence of submerged submarines by detecting faint distortions in the earth's magnetic field produced by ferrous objects. The aircraft needed to fly in a straight line, low over the water, in order for the device to work. The detection range of the device was between 120 and 250 meters. Therefore, a combination of an expert pilot flying low and "ideal conditions" gave a MAD aircraft the theoretical capability of detecting submarines at a depth of 300 feet. When the MAD picked up a target, it would illuminate a red light in the cockpit and automatically release an aluminum powder marker into the water: "The plane was then so maneuvered that the indicated position of the submarine was approached from four different directions, dropping aluminum powder markers each time. . . . In the center of the pattern of markers the submarine was supposed to be, waiting to be bombed."[76]

The IJN considered MAD-equipped aircraft to be one of their most effective ASW weapons, claiming that they sunk 15 submarines per month using this combination.[77] The device had difficulty distinguishing between submarines and sunken hulks, however, resulting in many attacks on shipwrecks. Furthermore, because of the device's short detection range, many more aircraft than were available were required for an effective sweep of convoy routes. While plans called for continuously sweeping 30 miles ahead of a convoy and a radar-equipped aircraft to monitor the rear of the convoy, the number of aircraft available never allowed for this. In practice, aircraft were able to search just ahead of convoys instead.[78]

The air convoy escort component of the IJN was deficient in both aircraft and detection equipment. The 901st Squadron, for example, which reached its maximum size in January 1945, consisted of just 170 mostly obsolescent planes, despite its authorized paper strength of 212 aircraft.[79] The shortfall was partly because of resource scarcity, but also due to the fact that the Combined Fleet staff was "short-sighted" in failing to allocate the resources it did have to maintain vital supply routes.[80] As the U.S. Bureau of Ships recorded, "By the end of the war only about one-third of the shore based anti-submarine planes had MAD, about one-third had radar, and only a very few were equipped with both. Anti-submarine planes were very seldom fitted with guns which accounts for the low incidence of strafing attacks on U.S. submarines."[81] The IJN used smaller fighter and light bomber aircraft converted for ASW use.[82]

Pilots for ASW duty were, as with the rest of the IJN, in short supply. Because of transportation difficulties within Japan, units did not train together. About a tenth of a unit's pilots would travel to training commands in the Home Islands, then travel back to their units to train the others. Furthermore, the pilots themselves objected to ASW duty, preferring assignments with "active combat."[83]

As Japan shrunk under the onslaught of the U.S. offensives during the war, the IJN's ASW aircraft suffered heavy attrition. Carrier air strikes on areas such as Formosa took a heavy toll on aircraft detailed for patrol and convoy escort. Casualties forced some units to withdraw from front-line service in spring of 1945.[84]

In the Battle of the Atlantic from 1941 to 1944, the escort carrier proved itself to be invaluable for Allied ASW efforts against the German U-boats. They were often converted from merchant ship hulls. They could only carry a few aircraft, but these provided air cover for convoys and greatly increased the effectiveness of hunter-killer groups. They were a particularly welcome addition to convoys while they sailed through the mid-Atlantic, out of range of land-based ASW aircraft. By comparison, the Pacific had longer lines of communication and fewer areas that could be covered by aircraft from island bases.[85] Japan only possessed four vessels of this type. The IJN made no attempt to use them as ASW flagships, and there were too few of them to make a difference in the overall Japanese commerce protection effort.[86]

The primary IJN aerial ASW weapon was the depth bomb. There were three types distinguished by their total weight: 60 kilograms, 180 kilograms, and 250 kilograms. Unlike the depth charges deployed by IJN surface craft, the fuses of the aerial weapons were time- rather than depth-triggered. In other words, the fuse measured the time from impact with the surface of the water, and the longer the time had been set before release, the greater the depth at which the weapon would detonate. The only way to change the time setting was before the weapon was loaded onto the aircraft, which severely limited the flexibility of the platform. The smaller bombs were carried on routine patrol, while the larger 250-kilogram weapon was only loaded when a definite contact with an enemy submarine had been established.[87]

Effective ASW required coordination of these various sensors and weapons with the platforms from which they were deployed. The most crucial element in that coordination was communication gear, another area in which the IJN was lacking. Unless ships had radio equipment on board, they had to communicate via signal flag, and communication between aircraft and surface ships was impossible. Coordination between IJN forces and Imperial Japanese Army (IJA) aircraft, which were sometimes engaged in convoy escort missions from bases such as New Guinea, was virtually impossible. The IJA pilots had little experience in flying over water, and the two services' communications equipment was incompatible. Therefore, even though both the IJN and the IJA saw fit to undertake air escort missions, there were no joint operations because of the lack of coordination capability.[88]

The most serious communication deficiency was not inter-service in nature, however. It was *intra*-service, between surface and air convoy escorts. Despite belonging to an air organization specifically for convoy escort, IJN aircraft were unable to communicate with surface vessels.[89] The radio sets developed for ship-to-aircraft communication never functioned properly.[90]

Thus, while both air and surface escorts were often present together when protecting a convoy, they could not relay information to one another or execute coordinated attacks on USN submarines once contact was made. At the convoy level, escorts frequently had inadequate communication facilities, and most merchant vessels had none.[91] This meant that simply coordinating the movement of the convoy itself more difficult than necessary, inhibiting timely and tactically sound responses to submarine attacks.

Organizational Deficiencies

The IJN's organization evolved considerably over the course of the Pacific War, but was deficient in two important ways. First, the responsibility of routing merchant traffic was split between several civilian and military organizations. Second, Japan lacked organizations for hunting submarines and organizing convoys.

Japan directed its merchant shipping through multiple organizations, which proved to be incredibly inefficient.[92] The routing of merchant traffic, which involved coordinating efficient transfer of material across the empire while simultaneously avoiding likely submarine operating areas, was the responsibility of the IJN, the IJA, and the Transportation and Commerce Department. The Munitions Department also had a "strong voice" in the matter of shipping routes. In short, those agencies that had cargo on merchant traffic had some say in what routes it would take. These military and civilian agencies would coordinate the routes in joint meetings, but the escort was left up to the Combined Fleet until late 1944. The fleet had no say about the routes of shipping; it only received requests for precious destroyers to be diverted from fleet duties to convoy protection.[93] Not surprisingly, because the commerce war was not the "decisive battle" in its view, the Combined Fleet command became "exasperated" with having to send its destroyers on missions in which it had little interest and no say as to when and where its resources would be committed.[94] Japan addressed this convoluted arrangement at the end of 1942 by establishing the half civilian–half military Shipping Control Board.[95]

From the outset of war, the IJN gave important convoys ASW protection, but they "did not expect aggressive and efficient submarine warfare to be advanced to their very shores."[96] The protection the IJN did provide was at first limited by local forces to the areas close to the convoy's destination. The primary open-ocean ASW measures undertaken were diverse routing and zigzagging (changing direction in order to disrupt torpedo attack). From the USN perspective, the zigzagging was of limited effectiveness, as a random turn was just as likely to produce a torpedo hit as it would a miss. Zigzagging also increased the time and distance a given ship had to travel, which gave the stalking submarine more opportunities to attack.[97]

Japan did not implement a general convoy system until early 1944.[98] Even with a convoy system in place, escort was left up to "area fleets," which resulted in inadequate allocation of convoy protection. Consequently, in November 1943, a command independent of the Combined Fleet, the Grand Escort Fleet, was formed. This organization consisted of three subordinate commands—the First and Second Escort Squadrons and the 901st Air Squadron. The First and Second Escort Squadrons had different geographic responsibilities. The First Escort Squadron became the First Escort Fleet; its areas of responsibility were the routes between Japan and the East Indies, the Philippines, and Palau. The Second Escort Squadron took the routes to the Marianas and the Caroline Islands. After the United States seized Saipan, the need for this squadron disappeared. Area fleets still retained escort duties for locales such as Singapore and Manila. The 901st Air Squadron was distributed across air bases astride imperial shipping routes with its headquarters located on the Home Islands.[99]

The IJN had no coherent doctrine for the operation of convoys. In general terms, the escort fleet had several subordinate commands, termed escort convoys. These were tactical units with no particular area of responsibility. Units assigned to the escort convoy varied with each mission; the escort fleet would retain senior officers and their staff to command convoys of merchant ships and available escort vessels as they were formed. Important convoys were assigned a captain or an admiral. The commander in chief of the escort fleet would determine the makeup of the convoy, in both merchants and escorts. The commander of each individual escort convoy was responsible for directing the routing of the merchant ships under his command.[100]

Tactically, by 1944, general practice had evolved so that convoys sailed with merchant ships sailing in a block formation, ringed by available escorts. In planning convoy routes, commanders placed heavy emphasis on evasive maneuvering and zigzagging. In other words, merchant ships steamed in two or three columns in the general shape of a square, while the escorts formed a circle around them.[101] The number of escorts available was always a serious handicap, and affected IJN convoy operations. Post-war USN analysis states that:

> In actual practice there were never enough escorts available. Convoys sometimes were held in port for lack of escorts, to be eventually sailed with whatever could be scraped up. Under these

> circumstances the promulgation of a doctrine was useless. The escort commander did the best he could with what he had available, and disposed of his escorts in accordance with his own ideas and hunches. Sometimes large convoys would be efficiently guarded by alert aggressive escorts. Sometimes a convoy would be trailed by a single escort vessel who could only harass the submarine after an attack had been made. Sometimes one flank of a convoy would be left completely unguarded. The behavior of the escort continued to be completely unpredictable.[102]

Indeed, even in more confined seas, lack of escorts affected operations. In the East Indies and New Guinea area, "no regular convoy routines were possible owing to the scarcity of escort vessels and their slowness."[103] The command responsible for escorting shipping within that large area, an example of an "area fleet," retained only one torpedo boat and 15 converted trawlers for escort duty in 1944. The trawlers had speeds of less than 15 knots and displaced less than 500 tons.[104]

Ironically, the severity of the U.S. submarine campaign and lack of adequate escort that forced Japan to form a separate command for escort duties in November 1943 also forced the re-subordination of the escorts to the Combined Fleet in August 1944. The Combined Escort Fleet needed assistance, and it was reasoned that the Combined Fleet destroyers would provide added firepower and numbers. This organizational change had the opposite effect, however. This was because after the reorganization, both the Combined Fleet as well as the Combined Escort Fleet could use the escort vessels previously reserved solely for shipping protection. The dwindling pool of available escorts now had to serve two masters. Japan was chronically short of escorts throughout the Pacific War, and this reorganization exacerbated the shortage.[105]

The IJN organized a unit that incorporated surface ships and ASW aircraft together for "hunter-killer" purposes in early 1945. The group consisted of five destroyer escort ships (the 102nd Surface Squadron), and twenty Navy fighters (the 934th Squadron), the latter equipped with both MAD and radar. The group patrolled in the East China Sea, between present-day Taiwan and Shanghai: "Both the planes and surface escorts were provided with voice radio but could communicate only for short distances and frequently were unable to communicate at all due to technical difficulties."[106]

Japanese intelligence in support of their ASW operations was not useful: "There is no evidence that the Japanese ever recovered anything of great value in anti-submarine warfare from their intelligence sources."[107] The USN knew that IJN intelligence was faulty because, among other things, in spring 1944 it had captured a report of Japan's Battle Lessons Investigation Committee, which analyzed the shipping situation from June 1943 to March 1944.[108] While the strategic position of U.S. submarine forces was fairly accurate, the IJN typically exaggerated the number of submarines that took part in any one incident.

More seriously, the USN reported that Japanese intelligence failed "to obtain any evidence upon which a proper evaluations [sic] of the effectiveness of their anti-submarine campaigning could be based. Japanese credited themselves with sinking about ten times the number of submarines that we actually lost."[109] As noted above, U.S. personnel noticed that the IJN persisted with ineffective ASW methods throughout the war. Because the IJN adapted to the submarine problem without knowledge of its mistakes or deficiencies, exacerbating its failure, this would suggest that the ability for accurate organizational introspection is important for successful adaptation.

A comparison with the Battle of the Atlantic demonstrates how instrumental accurate intelligence can be in ASW. In spring 1943, Allied convoy escorts sunk 41 German U-boats—30 percent of those at sea at the time. The Germans withdrew their boats from active confrontation with Allied convoys, but hunter-killer groups continued to harry them. Unlike the ineffective ASW sweeps of World War I (see Chapter 1), these groups were quite effective, because code-breakers in the Ultra program gave them accurate U-boat location information derived from decryption of German radio transmissions: "Their 'nose' for finding U-boats, seemingly uncanny at the time, can be attributed to Ultra."[110] U-boat casualties skyrocketed even as they avoided direct contact with Allied convoys; Allied intelligence negated German measures to reduce losses. In June 1943, Germany lost 17 U-boats; in July 1943, they lost 37.[111] Japan was unable to gather anything as accurate as the intelligence that the United States had gained from Ultra, and this was one of the many factors that made its ASW campaign less effective than that in the Atlantic, even though it faced fewer submarines.

Despite increasing experience with the U.S. mode of operation, Japan's effectiveness in countering it increased in 1943 but then reached a plateau: "Strangely enough, the rate of loss of submarines, directly due to enemy anti submarine efforts, was practically a constant for the years 1943, 1944, and 1945."[112] Furthermore,

> The early efforts of the Japanese to sink our submarines were largely counteracted by two things. (1) They set the depth charges

> too shallow. (2) They broke off the attack too soon. To some extent in both of these faults they persisted. They were always prone to accept the scantiest evidence that the submarine had been sunk and withdraw in triumph. Many a submarine owes its escape to this optimism. However, later enemy doctrine required persistent search and attack, and deeper setting of depth charges. After 1943 there appears to have been no major changes in the effectiveness of Japanese anti submarine methods.[113]

Despite experience facing U.S. submarines, the development of new technologies, the building of efficient ASW frigates, and reorganizing its fleet, the IJN did not increase its effectiveness in protecting Japanese commerce. With this in mind, we can now turn toward evaluating the IJN's ability to organize.

The Imperial Japanese Navy's Organizational Ability

We can now turn to evaluating the IJN's organizational ability in relation to the U.S. submarine threat. The IJN's resources—surface ships, aircraft, communications, and weapons—were all in short supply. There is no doubt that this had a significant impact on the organization's inability to combat the U.S. submarine menace. As explained in the following section, however, there were several organizational factors that made the resource situation worse than it would have been otherwise. Also, resources alone cannot account for the poor showing of the IJN's ASW and escort forces. The IJN was found lacking in the ability to form a dedicated organizational subunit (DOS) to address the U.S. submarine campaign in a timely manner. The IJN's level of organization relative to its adaptation requirement hindered its adaptation. In addition, it consistently failed to learn from its mistakes and evaluate the results of its efforts.

Resources

Japan had limited access to resources. The IJN's ASW and escort forces were of low priority in comparison to the Combined Fleet, and therefore suffered in terms of ships, aircraft, communications, detection equipment, and training.

Escorts were chronically in short supply. Often, convoys waited in port for lack of escorts. Also, many ships had to sail dangerous waters without escorts. Convoys that did sail often did so with inadequate escorts, which could only look on as U.S. submarines sank merchant vessels under their protection. When production of ASW frigates increased, there were still insufficient numbers of them.

The ASW aircraft situation was similar to that of the ships. Japan was stretched thin in terms of naval air power, so it could divert only a few obsolescent aircraft for ASW patrol and convoy escort duty. These numbers were further reduced by U.S. carrier raids on IJN bases and attrition in the air. Communications between the ships and ASW aircraft were problematic at best. Even though the IJN established a single command with both aircraft and surface ships for ASW, it never established satisfactory coordination between the two types of platforms because of inadequate equipment.

Another technological issue was that the IJN relied heavily upon the MAD; however, its faith in MAD was misplaced. MAD technology was in its infancy, so IJN devices had limited range. The aircraft that carried it also deployed inadequate weapons (e.g., obsolete depth charges), so that its effectiveness was mediocre at best. Furthermore, Japanese sonar technology lagged behind that of the United States. Perhaps most important, given heavy U.S. submarine use of surface cruising and surface search radar in its attacks, was the lack of radar on most Japanese escort vessels and aircraft. This allowed the U.S. submarines to use the cover of darkness and their own radar to sail ahead of merchant traffic and execute multiple attacks.

Finally, by the time the IJN realized the severity of its commerce protection problem, it had lost much of its capacity to train its personnel, and thus increase its expertise. Combined Fleet losses adversely affected ASW forces, because the IJN attempted to address those first. Finally, as evidenced by the uneven performance of escort forces, the IJN was unable to use the good practices of part of the force to improve the other parts.

It is clear that the necessary condition of adequate resources was not entirely present for the IJN. If the IJN had had enough resources, however, it could have made the U.S. submarine offensive a much more difficult endeavor. As in the BRN case in World War I, some of the above resource scarcity had its origins in the organization of the IJN's subunits.

The Struggle to Form a Dedicated Organizational Subunit

Much like the BRN at the outset of World War I, the IJN was focused upon the decisive clash of battleships as the central element of its naval strategy. Indeed, the BRN's organizational arrangement relative to the adaptation requirement, which is outlined in Chapter 2 of this volume, could easily be applied to the IJN as well.[114] There was an internal struggle within the IJN between the

Combined Fleet, which needed resources to prepare for its decisive battle, and the escort and ASW forces, which were concerned with supplying the both the Home Islands and Japan's outposts and fleets. Unlike the BRN, the battle fleet won the struggle at every turn. On one side, the adaptation requirement encompassed the Combined Fleet, as it possessed most of the ASW assets of the IJN in the form of fleet destroyers. The *kaibokan* were not yet in large-scale use or production. On the other side, there were a myriad of organizational subunits that had a say in commerce protection matters: because commerce protection fell under coastal defense, subunits with that responsibility would be the most important agencies; "area fleets" were expected to protect shipping within their geographic responsibilities; and finally, there was a constellation of subunits that all had a say in merchant traffic routing around the submarine threat. The two officers of the IJN staff assigned to address the commerce protection problem at the outset of war were so insignificant that they didn't even warrant a box on the organizational chart. While the USN started the war dividing its attention between warships and merchants, it was never very effective in targeting the former, and soon shifted focus to the latter. The result was an adaptation requirement that encompassed multiple subunits, but the most important subunits for active protection of commerce—area fleets and coastal defense commands—were never properly resourced. This arrangement is depicted in figure 6.1.

Besides the "broad stroke" conceptual division of the IJN into the Combined Fleet and "everything else," the IJN had some significant concrete organizational deficiencies. Part of this can be attributed to Japan's strategic

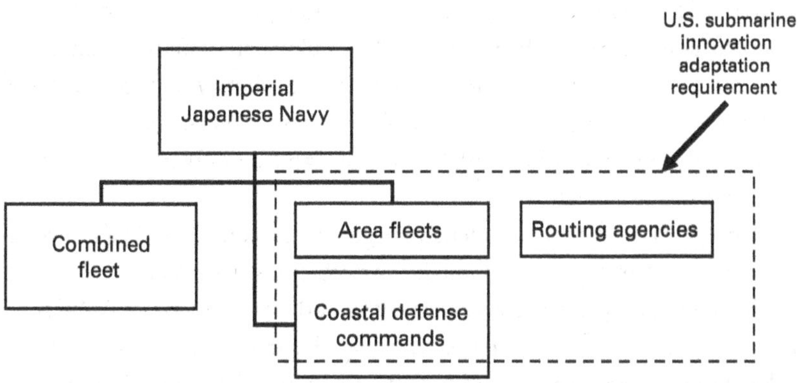

Figure 6.1
The Imperial Japanese Navy's initial organizational arrangement relative to the adaptation requirement.

emphasis on a short war, and the resulting lack of pre-war planning for commerce protection. Indeed, this strategic mindset in part accounts for why the U.S. submarine offensive, which was not novel on a global scale in any way, was "innovative" from the Japanese perspective.[115] The IJN was unprepared. The staff resources devoted to planning for commerce protection were so meager as to be almost nonexistent. The doubling of the number of officers devoted to such planning sounds impressive, until one realizes that there was only one officer assigned to this area prior to the war. This reflects the pre-war attitude that commerce protection was a "coastal defense" issue that could be handled by second-rate forces. It is all the more surprising given that one of the goals of launching the war in the first place was to ensure Japan's resource independence, and that those resources lay at the ends of very long sea-lanes of communications (SLOC).

Furthermore, the shipbuilding priorities of the IJN reflected the organizational dominance of the Combined Fleet. The *kaibokan*, the most effective class of ASW surface vessels that Japan produced, were short-changed in both displacement and numbers because of the demands on Japanese resources made by the building of the super-battleships. The prevailing organizational culture of the IJN reinforced these shipbuilding decisions. Like the BRN, it was dominated by offensive spirit. The professionally rewarding and exciting specialties lay in offensive naval aviation and the major combatant ships of the Combined Fleet. On the other hand, officers viewed ASW as unrewarding drudgework with limited career potential. While some thought had gone into commerce protection prior to the war, there was little motivation within the IJN to translate that thought into concrete action.

Once faced with a war on its commerce, however, the IJN failed to take action to address the problem. As the U.S. submarine offensive stagnated from lack of effective weapons and leadership, the IJN missed its opportunity to stamp it out while it retained control of its own waters. All of 1942 and much of 1943 passed without the IJN taking significant organizational steps to address this problem.

Once the U.S. submarine offensive began, Japan could have taken several measures to adapt to it, even though the problem was neglected before the war. It could have: 1) formed an escort command; 2) developed an ASW doctrine; 3) evaluated that doctrine's success based upon careful self-evaluation; and 4) modified its weapons and tactics based upon that evaluation. In fact, as 1942 ended, it had done none of these things. The IJN was lulled into a false sense of security by the ineffectiveness of the U.S. campaign. Because merchant shipbuilding kept pace with losses, the IJN viewed the problem as well in hand, rather than as the growing threat that it actually was. It ignored the fact that Japan was still losing merchant ships. In 1942, Japan still had a well-trained, well-equipped, and powerful Navy, which faced an adversary that was struggling with its own problems and hence vulnerable to effective countermeasures—but 1942 was a year of lost opportunities for the IJN.[116]

The years after 1942 were marked by further organizational mistakes. Even though an escort command was formed, no one command had responsibility for both routing of shipping and arranging escorts. "Area fleets" still retained their escort duties long after their strength had been whittled away. Furthermore, the IJN first improved its organization with the formation of the escort fleet, then degraded it by placing its resources at the disposal of the battle fleet. Therefore, in addition to the routing problem, the task of commerce protection was complicated even further by taking away some of the escort fleet's autonomy. Figure 6.2 represents this changing situation.

The IJN failed to develop a coherent ASW doctrine. Convoy tactics varied from convoy to convoy, depending on who its commander was. ASW tactics even varied from ship to ship. Some escorts were aggressive and effective, while others were incompetent. Even though resources were lacking, the promulgation of effective doctrine from those escorting forces that were

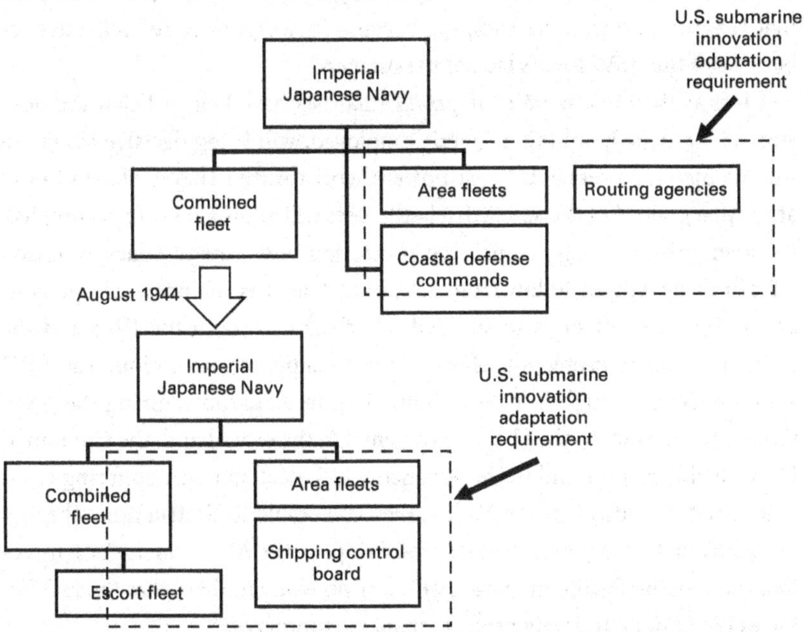

Figure 6.2
The Imperial Japanese Navy's mid-war organizational adaptation.

effective to those that were not could have increased the overall effectiveness of IJN escort forces. Also, as noted above, the IJN suffered from inadequate radar support. While certainly in part a resource issue, it should also be noted that the IJN persisted in relying on visual scanning as the primary method of detecting submarines. Radar was strictly a backup system; therefore, even when it was present, it was often underutilized.

Perhaps the most important organizational mistake, however, was the lack of accurate self-evaluation, because this led to the persistence of bad ASW practices by IJN forces. For example, the IJN failed to realize that its ships were making the rather simple mistake of setting their depth charges too shallow. Coupled with an inexplicable sense of over-optimism, this error allowed many U.S. submarines to escape unscathed. ASW forces assumed they were much more effective than they actually were, and no intelligence work contradicted that view. Furthermore, this demonstrates that resources were not the only shortcoming of ASW efforts. USN post-war analysis acknowledged that many of its submarines were: 1) found by IJN ASW forces; and 2) owed their survival to lack of persistence on the part of the IJN. So, even with rather mediocre equipment, the ASW forces could have been much more effective.

Lack of adequate intelligence support in part explains the IJN's failure to evaluate its ASW campaign. Without intelligence, it was unable to determine that U.S. submarines were escaping because its weapons were ineffective, or because of the ASW force's lack of persistence.

Just as the most fanatic air-power theorists have believed that the doctrine of strategic bombing, properly employed, will bring decisive results in war, staunch followers of U.S. sea-power theorist Alfred Thayer Mahan would argue that control of the sea with a battle fleet will allow a navy to accomplish its other missions (e.g., commerce protection and expeditionary warfare). All other capabilities follow from sea control, as it is the necessary precondition for other missions to succeed. On their face, both the IJN's and the BRN's commerce protection efforts seem to support this notion. The BRN kept the German High Seas Fleet bottled up in its harbors, giving the ASW forces time to address the U-boat problem. On the other hand, the Combined Fleet steadily lost ground to the advancing U.S. fleet, not only suffering casualties itself, but also leaving ASW forces vulnerable to destruction. The IJN lost many of its ASW aircraft and surface ships to the Allied carrier task forces that roamed the Pacific in 1944–1945. It is no wonder, then, that those ASW forces could not effectively protect Japanese commerce.

The relationship between the Combined Fleet and the ASW forces was more interactive than that rather simplistic view portrays, however. The failure of the Combined Fleet to win its decisive battle against the USN certainly did not help the ASW forces, but the failure of the ASW forces also in turn greatly inhibited the Combined Fleet. Lack of fuel prohibited the IJN from training replacement pilots, and greatly curtailed operations of its major surface units. Also, the lack of material available for Japanese industry meant that the IJN's strength dropped precipitously. These two factors meant that there were fewer ships and aircraft with which to protect Japan, and those that remained or were produced as replacements became progressively less effective as the war continued. Furthermore, the attrition of Japanese ASW forces by the U.S. carriers did not occur until 1944. The destruction of these forces came after the IJN had had its chance to stem the submarine war on Japanese commerce. In short, ASW forces did suffer from the IJN's loss of sea control, but that control was lost in part because of the failure of the ASW forces to protect Japan's SLOC.

In this case, the alternative theories of adaptation must be evaluated in a slightly different manner, as this is a case of adaptation failure. Rather than being responsible for a successful adaptation, did the lack of civilian intervention or competition better explain the Japanese failure in ASW than the lack of a DOS or top-level review panel (TLRP)?

The lack of civilian involvement in this case may indicate that outside intervention would have been necessary for adaptation. Left to its own devices, the Japanese military, including the IJN, developed doctrines and strategies that both ignored the possibility of a long war and set the nation on a collision course with a country that outclassed it in resources and industrial bases. The entirety of the Japanese war effort was military-dominated and separated from a realistic strategy. In the IJN, this manifested itself in the lack of pre-war planning for a commerce war, diverting shipbuilding resources away from ASW escorts to capital ships, and siphoning escorts from convoy duty to the Combined Fleet.

While the lack of civilian intervention in the IJN case is significant, it must also be noted that the IJN did eventually recognize the problem and try to correct it by implementing a convoy system without civilian involvement. The shortcomings in the adaptation were poor doctrine and execution, and lack of adequate ASW assets. There is little evidence that the civilian leadership saw the threat of U.S. submarines differently than the IJN. Even if it had, it is doubtful that this could have affected the faulty execution of the ASW campaign. If the Japanese political leadership had recognized the threat earlier than the IJN, it could have had an effect in successful adaptation if it had been able to override the IJN's shipbuilding priorities. Civilian intervention would have had to be very deep and comprehensive, however, given the nature of the IJN's failure in doctrine and execution of ASW. It is doubtful that Japan's political leadership could have accomplished this. The military's dominance over Japan's strategy also made the possibility of that kind of intervention exceedingly far-fetched.

The alternative competition explanation of adaptation could also pose a challenge to my top-down, internal theory of wartime adaptation. This is because: 1) the competition for resources between the Combined Fleet and the elements of the IJN responsible for commerce protection was never an even one; and 2) the Combined Fleet did not concern itself with finding a solution to the U.S. submarine problem. Therefore, competition did not really exist in the terms that Harvey Sapolsky lays out (divided workload, flexible jurisdictions between subunits) in his Fleet Ballistic Missile program case.[117] Because there was no competition, adaptation was weak and ineffectual.

If one considers, though, that at various points in the war there were two or three subunits responsible for ASW or convoy protection (area fleets, coastal defense commands, the escort fleet), the competition theory becomes somewhat weaker. They were all trying to accomplish the same basic task, so competition between them to accomplish that task in the best way should have produced more effective adaptation. The more effective subunits could have gained resources and influence as the IJN reinforced success. This did not occur, however; rather, each subunit foundered in its commerce protection task, and any successes or good practices remained isolated in the organizations and units in which they originated.

Conclusion

The results of this case study confirm that resources are necessary for successful adaptation, but not sufficient by themselves. Japan would likely have lost the Pacific War, given the disparity in resources, but its poor organizational response also contributed to accelerating the moment when it could no longer fight through the submarine blockade. Resource scarcities within Japan in general, and ASW forces in particular, were very important to the war's outcome. Organizational factors such as uncoordinated and separate commerce protection commands, lack of a coherent ASW doctrine, and failure to self-evaluate greatly exacerbated the IJN's resource weaknesses.

The IJN could have been much more effective if it had: 1) heeded the lessons of its shipping losses in 1942; 2) coordinated its ASW escort forces and shipping routes earlier; 3) put more shipbuilding resources into its *kaibokan* programs; and 4) more carefully analyzed its ASW efforts and corrected its incorrect view of their effectiveness. When one compares the mid-war adaptations of the BRN in World War I and the IJN in World War II, it becomes clear why Japan ultimately failed to protect its commerce. The constellation of organizations that Japan eventually arranged to deal with the U.S. submarine menace compares unfavorably in every aspect to the Antisubmarine Division (ASD) of the BRN.

Military innovation is indeed an interactive process between combatants. This case demonstrates that the decisiveness of an innovation is in large part influenced by how effective the target of that innovation is in responding to it. Besides the range, habitability, and radar of U.S. submarines, the USN's forces were not the most advanced or numerous in the world at the time. The USN had been designed for a completely different mission and equipped with extremely inefficient weapons. The United States expended just a tiny fraction of its overall war effort on the submarine war against Japanese commerce. Yet the U.S. submarine force was persistent and constantly improved its mode of operation. It is telling that when the pieces of an effective campaign came together at the end of 1943, it took only one year for this to become decisive. Conversely, the IJN stagnated, and became the victim of one of the most effective blockades in history.

7

Conclusion

THE CASE STUDIES in this volume have demonstrated three essential qualities of military innovation. First, it is an interactive process between the innovator and the adapter. Second, it is less likely to be decisive when the adapter is organizationally flexible. Third, it bestows a "first-mover" advantage that is often fleeting when facing an adversary with a strong ability to reorganize to meet the new threat. A quick review of the case studies is helpful in illustrating this. These conclusions also have important policy implications for the future, namely for organizations that must choose where to allocate scarce resources and find ways to react to unexpected events on the battlefield.

The Results of the Case Studies

The results of the case studies are shown below in table 7.1—a chart of innovation and innovator, adapter, organizational ability in terms of dedicated organizational subunit (DOS) and resources, and whether or not the adaptation was successful. Some adaptations came in phases; for example, the British Royal Navy (BRN) attempted an offensive strategy against U-boats, followed by a more effective organizational response. Similarly, the adaptation by the Imperial Japanese Navy (IJN) against U.S. Navy (USN) submarines came in three phases, none of them successful. These phases were uncovered during process-tracing, and they serve to increase the number of observations provided by the case studies.

The first case study of the BRN in World War I is perhaps the strongest example of the power of flexible organization in adaptation. Despite adequate resources available for adapting to the German U-boat threat and a tried-and-true solution available in the form of a convoy system, the BRN was unable to adapt successfully to the submarine menace through two years of war. The bulk of the BRN was offensively minded and viewed a convoy system as too defensive. Also, it persistently misinterpreted halts in the submarine campaign, brought on by U.S. diplomatic pressure on Germany, as the result of its own actions. Ultimately, the convoy system proved to be the adaptation that effectively blunted the German submarine offensive. Increasing losses, pressure from allies who were also losing ships, and the resources available

Table 7.1 The Results of the Case Studies: Successful and Unsuccessful Adaptations

Innovation (Innovator)	**Adapting Organization**	**Organizational Ability: Creation of DOS or TLRP?**	**Resources: Sufficient for Adaptation?**	**Adaptation Success?**
Submarine warfare on commerce (Imperial Germany)	British Royal Navy			
	Phase I: Offensive strategy	No	Yes	No
	Phase II: Organizational response	Yes	Yes	Yes
Improved anti-ship bombing (Imperial Japan)	U.S. Navy	Yes	Yes	Yes
Kamikaze (Imperial Japan)	U.S. Navy	Yes	Yes	Yes
Integrated air defense (North Vietnam)	U.S. Air Force			
	Phase I: Hunter-killer	No	Yes	No
	Phase II: Dempster Commission	Yes	Yes	Yes
Sea denial/ low-level attack (Argentina)	British Royal Navy	No	Yes	Yes
Submarine warfare on commerce (United States)	Imperial Japanese Navy			
	Phase I: 12/41–10/43	No	Yes	No
	Phase II: 11/43–7/44	Yes	No	No
	Phase III: 8/44–8/45	No	No	No

cannot account for the BRN's adoption of the convoy system. The creation of the Antisubmarine Division (ASD), and its assumption of all duties associated with combating the U-boat, was decisive in the adoption of the convoy system. The new subunit proved to be the only entity that was able to properly recognize the nature of the problem, decide on the best potential solution, and effectively advocate for that solution.

The second case study, the USN's adaptation to Japan's improved antiship bombing, demonstrates that having a dedicated organizational subunit (DOS) at the outset of a conflict is also beneficial to an adapter. With the raid on Pearl Harbor, the battleship fleet of the USN was out of commission; the contest between battleships and carriers for the role of capital ship was tilted inexorably in favor of the carrier. The USN, in the form of its carrier aviation, already had a well-developed DOS that could take sole responsibility for responding to the IJN's innovation. Not only did it respond effectively, but it also surpassed the initial innovation in striking power, durability, and operating efficiency.

The USN's further adaptation to the *kamikaze* innovation reinforces the idea that an existing DOS helps with adaptation. By late 1944, the USN had built its carrier force into the new focus of its battle fleet. When the *kamikazes* made their appearance, it was clear that the carrier aviation subunit would be instrumental in countering them. Together with improved technology, the United States was able to pour more resources into this subunit. The result was an adaptation using increased numbers, but good enough to prevent the Japanese innovation from being decisive.

In the third case study, the U.S. Air Force (USAF), like the BRN in World War I, started the war in Vietnam with a large base of resources but without a DOS to deal specifically with the threat of the air defenses of the North Vietnamese Navy (NVN). In the event of a war with the Soviet Union, the USAF reasoned that it could simply fly around that country's air defenses rather than attack them directly. Besides the logical fallacy of this (the Soviet Union could defend specific targets even if it could not defend all of its airspace), North Vietnam represented a different type of environment: smaller geography and fewer targets. Therefore, when the NVN used Soviet weapons during Operation Rolling Thunder, the USAF—specifically Tactical Air Command (TAC)—had to fly into the teeth of a sophisticated air defense system and was unprepared.

This case demonstrates that a comprehensive response to innovation need not arise out of a permanent DOS. Aircraft losses to surface-to-air missiles (SAMs) and anti-aircraft artillery (AAA) prompted the USAF to address the new problem. Although the Dempster Commission was not a DOS, it did represent an organizational response in the form of a top-level review panel (TLRP). Prior to the creation of this commission, the USAF's main subunits had not addressed the problem of SAMs in a coordinated or effective way. Strategic Air Command (SAC) possessed some countermeasures in the form of jamming pods, but its bombers practiced avoidance as their primary means of countering SAMs. With its lack of resources, TAC's tactical method amounted to avoidance only, which was ineffective due to the NVN's concentration of its defenses around targets that needed to be approached by TAC aircraft. The commission focused USAF efforts to develop and employ Electronic Countermeasures (ECM), anti-radiation missiles, and defense suppression aircraft ("Wild Weasels"), and allowed the USAF to work outside and across its normal internal divisions. The new agenda for adaptation allowed TAC to draw upon increased resources in the form of accelerated jamming pods and Wild Weasel aircraft development, personnel from SAC to fill vital electronic warfare officer billets, and anti-radiation weapons from the USN (e.g., Shrike and the STARM). The resources for adaptation were always present, but the USAF needed to develop a centralized approach to problem-solving in order for them to have an effect.

Furthermore, the commission developed the mission requirements for the Wild Weasel program, combining the task of both hunting and killing SAM radars into one aircraft. The Dempster Commission had wide-ranging and persistent power, speeding up research and development, allocating resources, and even exerting unofficial influence on defense contractors. Improving bombing accuracy at higher altitudes, however, was beyond its remit. U.S. fighter bombers continued to fly low-level bombing runs throughout the war, leaving them vulnerable to the NVN's AAA platforms. A more permanent DOS with authority to look at all aspects of the problem of suppression of enemy air defenses (SEAD) might have sped up the development of the laser-guided bomb, which did enable USAF aircraft to avoid AAA late in the war.

The fourth case study—the BRN and the Argentinian Navy in the Falklands War—illustrates the effects of focusing too much on one mission area prior to a war. With only months in which to adapt while simultaneously fighting the war, the BRN went to sea with a force much better prepared to hunt Soviet submarines in the North Atlantic than to fend off Argentinian anti-ship missiles and aircraft. Fortunately for the BRN, the process of shedding its power projection capability was not complete, and it was able to use its existing infrastructure and ship strength in the adaptation. The number of ships the BRN was able to send to the South Atlantic, along with its relatively large numbers of qualified personnel, gave it the ability to successfully retake the Falklands in the face of Argentinian opposition. This task proved to be costly, however. The adaptation was accomplished almost completely in the field in a piecemeal fashion, which in part accounts for the high casualties the BRN suffered. Only Argentinian mistakes prevented higher losses.

The fifth case study shows that failure is as instructive as success: Japan's failure to adapt to the U.S. submarine war on its commerce during World War II is a perfect example of this. A relatively small number of U.S. submarines was able to reduce the amount of raw materials and food imported into the Home Islands so substantially that the shortages irrevocably crippled the Japanese war effort. A cursory analysis might indicate that Japan did not have enough resources to commit to the submarine problem, but the root of the IJN's failure lay in both a shocking neglect of commerce protection in pre-war planning and a surprising lack of urgency through two years of war. While the BRN in World War I is a very strong example of the possibilities commensurate with the creation of a DOS, the IJN in World War II is a good example of the perils of the lack of that DOS. By the time the IJN attempted to centralize its anti-submarine warfare (ASW) effort, it had been sapped of its strength by the U.S. counteroffensive. Further exacerbating the problem, the IJN then reversed its organizational improvements in the ASW mission area to focus on its battle ships in the Combined Fleet.

What the Case Studies Prove

The strongest counterargument to the line of reason supported by these case studies is that sufficient resources are the dominant factor in effective adaptation. As table 7.2 demonstrates, however, organization was at least as important as resources. The relationship between creating a DOS and successful adaptation is stronger than that between possessing sufficient resources and adaptation success. This is supported by the case studies in two ways. First, successful adapters not only had sufficient resources, but also had (or formed) a DOS. Second, unsuccessful adapters more often than not had sufficient resources available but failed to use them properly because they did not have or create a DOS. In other words, the creation of a DOS was more decisive than resources.

Table 7.2 depicts the results of the case studies in terms of the dependent variable—successful adaptation. There were five cases or case phases of adaptation success: the BRN's organizational response to the U-boat (Phase II); the USN's adaptations to the IJN's improved anti-ship bombing and the *kamikaze*;

Table 7.2 The Results of the Case Studies: Phases of Adaptation Success

Innovation (Innovator)	**Adapting Organization—Phase**	**DOS or TLRP?**	**Resources?**	**Adaptation Success?**
Submarine warfare on commerce (Germany)	British Royal Navy Phase I: Offensive strategy	No	Yes	No
Integrated air defense (North Vietnam)	U. S. Air Force Phase I: Hunter-killer	No	Yes	No
Submarine warfare on commerce (United States)	Imperial Japanese Navy Phase I: 12/41–10/43	No	Yes	No
Submarine warfare on commerce (United States)	Imperial Japanese Navy Phase II: 11/43–7/44	Yes	No	No
Submarine warfare on commerce (United States)	Imperial Japanese Navy Phase III: 8/44–8/45	No	No	No
Submarine warfare on commerce (Germany)	British Royal Navy Phase II: Organizational response	Yes	Yes	Yes
Improved anti-ship bombing (Japan)	U.S. Navy	Yes	Yes	Yes
Kamikaze (Japan)	U.S. Navy	Yes	Yes	Yes
Integrated air defense (North Vietnam)	U.S. Air Force Phase II: Dempster Commission	Yes	Yes	Yes
Sea denial/low-level attack (Argentina)	British Royal Navy	No	Yes	Yes

the USAF's adaptation to the NVN's integrated air defenses (Phase II); and the BRN's adaptation to the Argentinian Navy's low-level air attacks. In four out of five cases or case phases, both a DOS and sufficient resources were present. In only one case—that of the BRN during the Falklands War—did the adapting organization lack a DOS. As noted in the case study, however, the BRN suffered high casualties because of the *ad hoc* nature of its adaptations. These cases of adaptation success indicate that successful organizations have both resources and a DOS.

There were five cases or case phases of adaptation failure: Phase I of the BRN's adaptation to the U-boat; Phase I of the USAF's adaptation to the NVN's air defenses; and Phases I, II, and III of the IJN's adaptation to the U.S. submarine offensive. In one of these, Phase III of the IJN's adaptation to the U.S. submarine campaign, neither a DOS nor adequate resources were present. Successful adaptation in this case was extremely unlikely. In three out of the remaining four cases, there were sufficient resources for adaptation, but not a DOS. The BRN early in World War I, the USAF in Vietnam, and the IJN early in World War II all had sufficient resources but lacked a DOS. In the IJN's case, there was a DOS but not sufficient resources for adaptation. Therefore, the weight of the evidence indicates that resources are necessary for adaptation, but not sufficient.

The cases show that when an organization forms a DOS for adaptation purposes, it will probably be successful, given adequate resources. The mere presence of those resources is not decisive by itself. Organizations with adequate resources for adaptation failed to adapt successfully in three instances, and in only one of these cases did the lack of resources prove more decisive than organizational inflexibility: that of the IJN's ASW from November 1943–July 1944. Furthermore, only one organization without a DOS throughout a conflict—the BRN in the Falklands War—was able to successfully adapt. The adaptation process proved to be relatively costly, given the size of the conflict. The fact that out of five successes, success was achieved in only one case without an organizational response strongly supports the notion that resources are necessary for successful adaptation, but not sufficient.

Examining Other Possible Explanations About the Causes of Adaptation

In examining how militaries successfully adapt, my theory also establishes that more often than not, adaptation is imposed from within the organization and in a top-down fashion. There are alternative explanations about the sources of adaptation, as the innovation literature contains alternative theories about the sources of innovation. The relevant possibilities here are civilian intervention and competition between and among military subunits. In each of the case studies, I have evaluated these alternative explanations. They provide some challenges to my theory in some of the case studies; however, process-tracing shows that internal, top-down organizational measures, rather than civilian intervention or bureaucratic competition, were more decisive in enabling organizations to think about the new problem, develop a solution, and implement that solution.

Civilian intervention is the weaker of the two main alternative theories about wartime adaptation in these case studies. It was at its strongest in the case of the BRN in World War I; in that case, the U-boat problem was serious enough that there was significant pressure by civilians on the BRN to adopt the convoy system. This case study reveals, however, that internal reorganization—the creation of the ASD—enabled the BRN to finally recognize that past measures were not working and support the idea of the convoy. For the USN in World War II, civilian intervention was not a factor in its adaptations; it focused on carrier aviation because that was the instrument available. In the case of the USAF in Vietnam, civilian intervention was actually counterproductive, because the retaliation strategy only resulted in more casualties from the NVN's SAMs. In the Falklands case, the civilian actions that most affected the BRN's adaptation to the Argentinian innovation were convincing the French to stop export of Exocet anti-ship missiles and allowing the BRN to neutralize the Argentinian surface threat. The former was a diplomatic accomplishment, and the latter was an adjustment in the rules of engagement (the actual idea of sinking the *Belgrano* came from within the BRN). Therefore, calling these measures "adaptations" is problematic at best. Rather, they were skillful diplomacy and simple war-fighting, respectively. In the case of the IJN and USN submarines in World War II, there was no appreciable civilian intervention. In order to reverse the IJN's failures in commerce protection, however, any civilian intervention would have had to be comprehensive and far-reaching, ranging from shipbuilding priorities to allocation of ASW units within the fleet to correcting faulty doctrine. It is doubtful that the civilian Japanese leadership could have accomplished this.

Competition between and among military subunits also had, at best, varying effects in the case studies. In the case of the BRN in World War I, there was no meaningful competition between the battle fleet and other subunits. The prevailing view within the BRN was that the bulk of its combat power resided in the battle fleet and it was the ultimate guarantor of Great Britain's security. Other parts of the fleet—e.g., destroyers and other anti-submarine forces—played supporting roles to the battleships. For the USN in World War II, no meaningful internal competition took place either. The IJN eliminated the U.S. battle fleet as an effective competitor to the carriers. In the Vietnam case, the USN shared much of its SEAD technology with the USAF, and the services' respective responsibilities in Operation Rolling Thunder were clearly delineated. This lessens the impact of the competitive explanation of successful adaptation. In the Falklands case, the BRN was virtually the only service responding to the Argentinian innovation. Furthermore, it was the loser of a long-term competition for resources with the Royal Air Force (RAF). Finally, in the case of Japan and the U.S. submarine offensive, there was no meaningful competition over solving the problem in the IJN.

Implications and Recommendations

The argument presented in this book has many policy implications, especially in the area of organizational choices. First and foremost, a "first-mover" advantage granted by innovation does not in any way guarantee victory. Relying solely upon the innovative use of military forces to achieve victory would be a mistake. Other factors—especially the ability of the opponent to organize to meet the new threat—carry more weight in determining which side emerges victorious. In trying to determine the potential effects of an innovative mode of operation, an organization must consider its likely opponent and the vulnerabilities of the innovation, because a competent opponent will tend to exploit the latter. It is dangerous to assume that an opponent can not or will not adapt.

Second, the case studies demonstrate that military organizations need to be flexible. This statement is true of all organizations that are engaged in competitive endeavors; their leaders must be able to think on their feet and adjust their actions to changing conditions. Military organizations pursuing a first-mover advantage through innovation, however, must also consider both their adversary's possible adaptations and ways to preserve their innovative gains. As some have argued, militaries might be better served by maintaining organizational flexibility, rather than by trying to predict their adversary's behavior.[1] This may work well in land war, but the maritime domain poses different challenges. The technological nature of naval forces necessitates large and time-consuming investments in new machines, as well as technological expertise. Therefore, the trajectories of maritime weapons-development programs cannot turn on a dime. In all but the longest of wars, conflicts in the maritime domain will be "come as you are" affairs, with little time to make far-reaching adaptations, especially in the technological realm.

First-Mover Advantage

The case studies demonstrate that first-mover advantages generated by innovation disappear far too rapidly in wartime for militaries to depend on them for victory, especially if they intend to challenge a competent opponent. In a general sense, this means that a more profitable expenditure of scarce resources would be to build into a force structure the necessary "slack" for adaptation and exploitation of new knowledge or experience. Additionally, one way to hedge against a canny adversary may be to use war games and other analytical events to "game out" how new technology could be used, consider likely counters to that technology, and develop alternative uses that may mitigate the adversary's adaptations.

The case of the USN and IJN innovation is particularly instructive here. The IJN built a very effective carrier air arm prior to World War II; however, its obsessive focus on achieving qualitative superiority over the USN led it down the path of creating what was essentially a one-use weapon. In almost every area, from the number of trained pilots to the durability of its ships and aircraft, Japanese carrier aviation was extremely vulnerable to a counterblow if it failed to defeat its enemy with the first strike. In its quest for qualitative superiority, the IJN "eschewed the good for the perfect," and achieved neither. It is fair to say that the IJN could never win a contest with the USN in terms of material resources, but it worsened its own situation by neglecting the training of pilots before the war and failing to ramp up the training program once war began. Furthermore, the IJN's carrier aviation was remarkably stagnant after the war began, failing to develop more robust ships and aircraft or to change the composition of its air wings to help increase the survivability of its carriers against the rapidly developing U.S. carrier aviation capability.

Flexibility

This volume also makes a strong case for organizational flexibility: what characteristics make organizations flexible, how a DOS aids flexibility, and how specialization in mission can inhibit it.

In general terms, these cases suggest that it is vital for military organizations to ensure that they have: 1) mechanisms (e.g., good intelligence and analysis) for recognizing unexpected problems; 2) freedom to create new subunits; and 3) spare resources to allocate to new subunits. This prescription also implies that policymakers and military leadership should not commit to war a military organization that does not have some spare capacity in resources, primarily expert personnel. The IJN is again instructive; resource shortages continuously hampered its efforts to adjust to the U.S. submarine offensive.

The historical case studies recounted here point to a number of reasons why an organization might create a DOS: 1) leadership identifies some new task, imposed by the innovation, for which no existing subunit exists; 2) the organization suffers a defeat just short of catastrophic and this painful event nonetheless leaves enough of the organization to recognize the defeat and reassess, leading to organizational change; 3) one or more of the subunits competing for the innovation task are eliminated, often in the process of defeat; and 4) in a more gradual process, the organization recognizes that it is suffering casualties that push it toward failing in its objectives.

Specialization, or fixation upon one mission at the expense of others, seems to limit flexibility, which in turn negatively impacts adaptability. The case studies in this volume provide several examples of organizations that made choices to specialize in specific areas prior to the particular conflict under examination, and paid the price in wartime. First, the BRN, for example, focused its pre–World War I efforts on maintaining a dominant battle fleet. In this it was quite successful, but this was to the detriment of other important capabilities, and the price of this was that it neglected what turned out to be the decisive mission—commerce protection. Second, it is worth noting that at the time of this writing, the USN had been operating in a virtually uncontested environment for close to three decades; with the collapse of the Soviet Union, no other power could rival the United States in the size or capabilities of its navy. In that time, the USN specialized in supporting operations ashore rather than fighting potential rival navies at sea. While there were no plausible rivals for much of that period, the growth of the People's Republic of China Army Navy in size and capability in recent years has led to a renewed focus on war-fighting at sea. The USN, however, has much work to do to make up for the decades of neglect.

As noted above, IJN carrier aviation stagnated in the face of U.S. adaptation. It is worth emphasizing this again in this respect: the IJN was so focused on its version of carrier warfare that it could not change its carrier force's staying power to better combat the adaptations of its U.S. counterpart, nor could it adapt in other areas, such as the war on Japanese commerce. Short-war focus dominated almost every aspect of Japan's planning. Therefore, IJN carrier aviation was still in a strong auxiliary role well into the war, and Japan neglected to secure its trade routes, a medium-term and long-war concern. Strategies for quick victory that seek to mitigate risks associated with novel methods of warfare, or strategic vulnerabilities, are not reliable ones, especially if the opponent is a great power.

The USAF's SAC and TAC specialized in nuclear weapons delivery against the Soviet Union in the early 1960s. This specialization made TAC particularly vulnerable to the NVN's integrated air defense (IAD) during Operation Rolling Thunder. The USAF degraded the effectiveness of the SAM component relatively quickly, but AAA always remained a problem. Accurate bombing from medium and high altitudes could have addressed AAA by avoiding it altogether, but the capability to do this did not exist until late in the conflict, with the widespread use of the laser-guided bomb (LGB) during Operation Linebacker I. Presumably, the USAF could have developed the LGB sooner if it had not favored nuclear weapons (with their reduced need for accuracy) so strongly.

Ad hoc organizational arrangements are less effective but still make a positive impact on the ability to adapt or innovate. Concerted, organized efforts to solve problems can yield unexpected benefits. Many technologies vital to counter SAMs, e.g., electronics and missiles, were imported from the USN after the Dempster Commission. Would this transmission of technology across service lines have occurred without the commission to chart a way forward and coordinate this interaction?

Innovation and organizational change should be undertaken with the core missions of the organization in mind. The Falklands War caught the BRN at the tail end of a transition from global exercise of sea power to a role tailored for ASW in the North Atlantic in support of NATO. This transition neglected two things: 1) Great Britain still had overseas possessions to defend; and 2) air defense and the fleet air arm were still vital to the BRN's mission, even if it was confined to the North Atlantic. The BRN's difficulties in the South Atlantic demonstrate that divesting in too much capability and core competencies carries significant risk, particularly if the strategic environment is such that the organization may need the capability that it has shed or is in the process of shedding. The BRN was fortunate that its specialization in a NATO-oriented ASW mission had not progressed further at the time of the Falklands War. Had Argentina merely waited a few more weeks before occupying the Falklands, the BRN would have shed enough of its expeditionary and aviation capabilities to make its operations impossible. The nature of this conflict demonstrated that the BRN's ships were out of their element. Without the RAF to provide air cover and Airborne Early Warning (AEW), the BRN's commander was forced to take *ad hoc* measures to fill these gaps in capability. This was not simply a matter of the number of hulls. If the BRN's ships had possessed more Sea Wolf systems, and if the carriers had retained an AEW capability, the BRN would have reduced its losses dramatically.

A Way Forward: Areas of Further Research

One of the primary goals of this project—evaluating innovation's effect on war outcomes—is to inform the choices military organizations must make when allocating resources. In furtherance of that goal, future research should proceed in a two-step process. First, the theory needs to be further tested in the arenas of air and land combat. Second, guerilla war and counterinsurgency are obvious avenues of future inquiry.

This project has focused primarily on navies, with one air warfare case. To determine the general applicability of the hypothesis—that innovation's effect on war outcomes depends upon the adversary's ability to adapt—more case studies in other areas of warfare are needed. If we return to the universe of cases of wartime innovations, several obvious choices emerge. For example, the *levee en masse* and the corps system of the Napoleonic wars, the various infantry innovations of Imperial Germany, and the *Wehrmacht*'s innovations during World War II stand out. Cyber conflict has become increasingly sophisticated and organized. To what extent have the various types of organizations developed for this new arena—U.S. Cyber Command, for example—enabled militaries to adapt to unexpected or novel cyber threats? Has the use of unmanned aerial vehicles (UAVs) for reconnaissance and combat missions prompted militaries to change their organizations, and to what extent, if any, have those changes improved operations?

Alternatively, the Arab-Israeli wars of the twentieth century, especially the 1973 Yom Kippur War, are potentially informative case studies. The Israel Defense Forces (IDF) faced relatively innovative use of military hardware, particularly by Egypt, during that conflict. Egypt found ways to blunt two of the IDF's primary strengths: its armored forces on land, and its air force in the sky. Not only would this conflict provide useful details about the innovator-adapter relationship in the areas of air and land combat, but it would also test the thesis in wars between smaller powers. Furthermore, it would enhance the comparative analysis of the project by extending it to non-Western militaries.

Furthermore, innovation need not be the exclusive domain of organized militaries. Is the operational level of analysis appropriate to analyze innovation in asymmetric warfare? What kinds of military organizations are successful at counterinsurgency operations or asymmetric warfare? Would the DOS have the same effect as it does in sea and air combat? Does flexibility matter in low-intensity conflict? If it does, are there ways to institute flexibility in different types of organizations? The answers to these questions could have important lessons for military organizations that must plan for and address changes and uncertainty in the future of warfare.

Finally, the apparent reemergence of Great Power competition argues for more intense study of the maritime domain. Policymakers and analysts seem to have realized that Great Power competition does not stop, even when the largest economic and military power has forgotten about it. Adversaries of the United States have been innovating—with the aim of eroding U.S. military superiority—for more than two decades. The renewed possibility of Great Power war means that the future potential for multiple complex adaptation challenges is greater than ever. The implications for the balances of power between the United States and the People's Republic of China, and the United States and Russia, as well as the potential emergence and growth of atypical and asymmetric adversaries, clearly illustrate the importance of further research on military innovation and adaptation in war.

Notes

Introduction

1. Morton Halperin discusses "organizational essence" and its effects (positive and negative). How an organization defines its essence has powerful effects, namely in how it pursues tasks that reinforce its essence and shuns ones that do not. The U.S. Air Force aggressively pursued independence from the U.S. Army, but was also extremely reluctant to support the Close Air Support (CAS) mission, to the detriment of the U.S. Army in the post–World War II and Korean War periods. CAS was outside of the USAF's perception of its essence. See Morton H. Halperin, *Bureaucratic Politics and Foreign Policy* (Washington, D.C.: Brookings Institution Press, 1974), pp. 39–40 and 43–46.
2. I have based my methods on those laid out in works such as Stephen Van Evera, *Guide to Methods for Students of Political Science* (Ithaca: Cornell University Press, 1997); Alexander L. George and Andrew Bennett, *Case Studies and Theory Development in the Social Sciences* (Cambridge, Mass.: MIT Press, 2005); David Collier, "Understanding Process Tracing," *PS: Political Science and Politics*, Vol. 44, No. 4 (2011), pp. 823–830; and Jacob Ricks and Amy Liu, "Process Tracing: A Practical Guide," *PS: Political Science and Politics*, Vol. 51, No. 4 (October 2018), pp. 842–846.
3. The U.S. Joint Chiefs of Staff (JCS) provides one useful definition of the levels of war: strategic, operational, and tactical. The strategic level involves using elements of national power as means in pursuit of national-level ends (objectives): "Joint doctrine focuses [tactics] on planning and executing battles, engagements," while "the operational level of warfare links the tactical employment of forces to national strategic objectives." See Chairman of the JCS, "Joint Publication 3-0: Joint Operations," October 22, 2018, I-12–14, accessed at: https://www.jcs.mil/Portals/36/Documents/Doctrine/pubs/jp3_0ch1.pdf.
4. War Plan Orange was the U.S. plan for defeating Japan. It was part of the "Rainbow" series of plans, which used a color code for each potential adversary. For a history of the design, evolution, and execution of Plan Orange, see Edward S. Miller, *War Plan Orange: The U.S. Strategy to Defeat Japan, 1897–1945* (Annapolis: Naval Institute Press, 1991).
5. Operation Rolling Thunder was the Lyndon Johnson administration's coercive air campaign against North Vietnam, conducted from 1965–1968. Operation Linebacker I was the Richard Nixon administration's 1972 air campaign to blunt North Vietnam's conventional Easter Offensive against South Vietnam, while Linebacker II targeted Hanoi and Haiphong Harbor in North Vietnam with the aim of forcing that country back to peace talks in Paris.

6. Meir Finkel in particular highlights the importance of flexibility among military organizations, because accurate prediction of what future warfare might look like is often a futile exercise. See Meir Finkel, *On Flexibility: Recovery from Technological and Doctrinal Surprise on the Battlefield*, Moshe Tlamim, trans. (Stanford: Stanford University Press, 2011).

Chapter 1

1. For a survey of military innovation literature, a good place to start is Adam Grissom, "The Future of Military Innovation Studies," *Journal of Strategic Studies*, Vol. 29, No. 5 (October 2006), pp. 905–934. For a critical assessment of the state of the field ten years after Grissom's work, see Stuart Griffin, "Military Innovation Studies: Multidisciplinary or Lacking Discipline?" *Journal of Strategic Studies*, Vol. 40, Nos. 1–2 (2017), pp. 196–224.
2. Exploring why military organizations pursue unsuccessful or ineffectual innovations may be a useful line of inquiry, however.
3. Stephen Peter Rosen, *Winning the Next War: Innovation and the Modern Military* (Ithaca: Cornell University Press, 1991). One of Rosen's primary theoretical goals is to determine the source of military innovation. While I am using his definition, I do not necessarily accept his theory on the source of innovation.
4. Matthew Evangelista, *Innovation and the Arms Race: How the United States and the Soviet Union Develop New Military Technologies* (Ithaca: Cornell University Press, 1988), p. 12.
5. Owen R. Coté, Jr., "The Politics of Innovative Military Doctrine: The U.S. Navy and Fleet Ballistic Missiles," Ph.D. dissertation, Massachusetts Institute of Technology, 1996, p. 9.
6. Barry R. Posen, *The Sources of Military Doctrine: France, Britain, and Germany Between the World Wars* (Ithaca: Cornell University Press, 1984), pp. 130–132.
7. Ibid., pp. 167, 174.
8. Ibid., pp. 190–212.
9. Rosen, *Winning the Next War.*
10. Richard Overy, *Why the Allies Won* (New York: W.W. Norton and Company, 1995), p. 62.
11. Ibid., p. 100.
12. Keir A. Lieber, *War and the Engineers: The Primacy of Politics Over Technology* (Ithaca: Cornell University Press, 2005), pp. 30–33.
13. Ibid., p. 29.
14. Ibid., p. 3.
15. George Raudzens, "War Winning Weapons: The Measurement of Technological Determinism in Military History," *Journal of Military History*, Vol. 54, No.4 (October 1990), pp. 403–434.

16. Kenneth Waltz, *Theory of International Politics* (Boston: McGraw-Hill, 1979), pp. 96, 181.
17. George William McNeill, *The Pursuit of Power: Technology, Armed Force, and Society Since A.D. 1000* (Chicago: Chicago University Press, 1982), pp. 123–143.
18. For example, Tor Bukkvoll examines Russian Special Operations Forces innovation as a top-down process in an authoritarian regime context. See Tor Bukkvoll, "Military Innovation Under Authoritarian Government—The Case of Russian Special Operations Forces," *Journal of Strategic Studies*, Vol. 38, No. 5 (2015), pp. 602–625. Raphael Marcus's Israel case posits a combination of bottom-up and top-down sources of innovation. See Raphael Marcus, "Military Innovation and Tactical Adaptation in the Israel-Hizballah Conflict: The Institutionalization of Lesson-Learning in the IDF," *Journal of Strategic Studies*, Vol. 38, No. 4 (2015), pp. 500–528. In arguing that Singapore's modernization constituted innovation, Evan Laksmana theorizes that top-down factors were instrumental. See Evan Laksmana, "Threats and Civil-Military Relations: Explaining Singapore's 'Trickle-Down' Military Innovation," *Defense and Security Analysis*, Vol. 33, No. 4 (2017), pp. 347–365. Innovation can also travel across operational and tactical organization within an army. See Robert T. Foley, "A Case Study in Horizontal Military Innovation: The German Army, 1916–1918," *Journal of Strategic Studies*, Vol. 35, No. 6 (December 2012), pp. 799–827. Andrew Hill identifies cultural factors, namely the alignment of an innovation with an organization's "cultural concept of the ideal combatant," as instrumental in the organization "developing and implementing the change." See Andrew Hill, "Military Innovation and Military Culture," *Parameters*, Vol. 45, No. 1 (Spring 2015), pp. 85–98. The U.S. Air Force's innovative doctrine resulted from interservice cooperation with the U.S. Army, due to a special confluence of budgetary austerity and the Soviet threat. See Phil Haun, "Peacetime Military Innovation Through Inter-Service Cooperation: The Unique Case of the U.S. Air Force and Battlefield Air Interdiction," *Journal of Strategic Studies*, DOI: https://doi.org/10.1080/01402390.2018.1557053, pp. 1–26. Finally, in their study of innovation since the Cold War, Theo Farrell, Sten Rynning, and Terry Terriff eschew testing a theory and focus instead on a comparative study of how U.S., British, and French armies changed over two decades. See Theo Farrell, Sten Rynning, and Terry Terriff, *Transforming Military Power Since the Cold War: Britain, France, and the United States, 1991–2012* (Cambridge: Cambridge University Press, 2013), pp. 8–10.
19. Leslie C. Eliason and Emily O. Goldman, "Introduction: Theoretical and Comparative Perspectives on Innovation and Diffusion," in Emily O. Goldman and Leslie C. Eliason, eds. *The Diffusion of Military Technology and Ideas* (Stanford: Stanford University Press, 2003), p. 1.
20. Ibid., pp. 2–6.
21. During the preparations for the Iraq War, administration planners (led by Secretary of Defense Donald Rumsfeld) continually pressed Central Command to "do more with less." Increased weapon precision and lethality, better intelligence, and faster communications fostered by the RMA would obviate the need for a large land

army in the style of Desert Storm. For an in-depth discussion of the U.S. planning for the Iraq War (and the influence on that process on the Afghan campaign), see Michael Gordon and Bernard Trainor, *Cobra II: The Inside Story of the Invasion and Occupation of Iraq* (New York: Pantheon Books, 2006). For an in-depth analysis of innovation in U.S. Army doctrine between the Vietnam and Iraq Wars and an alternative theory of the sources of military innovation, see Peter Campbell, *Military Realism: The Logic and Limits of Force and Innovation in the U.S. Army* (Columbia: University of Missouri Press, 2019). Campbell theorizes that innovation springs from classical realist tendencies among senior military leaders, in contrast to Posen's civilian intervention model.

22. This has been persuasively argued by Christopher Layne, "This Time It's Real: The End of Unipolarity and the 'Pax Americana.'" *International Studies Quarterly*, Vol. 56, No. 1 (2012), http://www.jstor.org/stable/41409832, pp. 203–213. Since that piece was published, other notable indicators have been the growth of the military in the People's Republic of China and changing balances of power in Europe and the Middle East.

23. The term "Great Power competition" was codified into U.S. security policy in the 2017 *National Security Strategy of the United States of America*, https://trumpwhitehouse.archives.gov/wp-content/uploads/2017/12/NSS-Final-12-18-2017-0905.pdf. For a contemporary discussion of the popular emergence of the term, see Uri Friedman, "The New Concept Everyone in Washington Is Talking About," *The Atlantic*, August 6, 2019, https://www.theatlantic.com/politics/archive/2019/08/what-genesis-great-power-competition/595405/.

24. James Q. Wilson, *Bureaucracy: What Government Agencies Do and Why They Do It* (New York: Basic Books, 1989), p. 222.

25. Williamson Murray, *Military Adaptation in War: With Fear of Change* (Cambridge: Cambridge University Press, 2011), p. 308.

26. To some extent, this conflicts with much of the literature that uses the concept of "adaptation" as a less in-depth form of innovation. In his analysis of wartime innovation, James Russell explicitly links innovation and adaptation as degrees of the same phenomenon: "Tactical adaptation can serve as a waystation along the route toward more comprehensive innovation." Similarly, Theo Farrell defines adaptation as "change to strategy, force generation, and/or military plans and operations, undertaken in response to operational challenges and campaign pressures"; however, "military innovation implies a greater degree of novelty and disruptive organizational change than adaptation." Theo Farrell, "Introduction," in Theo Farrell, Frans Osinga, and James A. Russell, eds., *Military Adaptation in Afghanistan* (Stanford: Stanford University Press, 2013), pp. 2, 7. Matthew Ford also notes that adaptation is "adjusting existing means and methods." Matthew Ford, *Weapon of Choice: Small Arms and the Culture of Military Innovation* (Oxford: Oxford University Press, 2017), p. 17.

27. Organizations involve a "system of coordination," a "degree of centralization," and a concept of "how to perform a critical task" in support of a mission. Wilson, *Bureaucracy*, pp. 24–25.

28. Trent Hone describes how the USN used boards—essentially TLRPs—to address a host of issues including gunnery inaccuracy, doctrine and tactics, and ship design. See Trent Hone, *Learning War: The Evolution of Fighting Doctrine in the U.S. Navy, 1898–1945* (Annapolis: Naval Institute Press, 2018), pp. 76–91, 327–328.
29. Morton H. Halperin, *Bureaucratic Politics and Foreign Policy* (Washington, D.C.: Brookings Institution Press, 1974), p. 26.
30. Ibid., p. 27.
31. Ibid., p. 40.
32. Samuel P. Huntington, *The Soldier and the State: The Theory and Politics of Civil-Military Relations* (Cambridge: Harvard University Press, 1957), p. 67.
33. Ibid., pp. 43–46.
34. Indeed, Wilson notes that while "missions" are often quite vague, the organization defines for itself its "critical tasks" in support of those missions, which determine what activities it actually does. Wilson, *Bureaucracy*, p. 25.
35. Anthony Downs, *Inside Bureaucracy* (Prospect Heights, Illinois: Waveland Press, 1994), p. 212.
36. Ibid., pp. 215–216.
37. U.S. Department of the Navy, Naval Historical Division, *U.S. Naval Administrative Histories of World War II Number 89: An Administrative History of the Bureau of Ships in World War II*, Vol. I, Naval Historical Center Archives, Washington Navy Yard, Washington, D.C., p. 21.
38. Only "line officers" were eligible to command at sea. During this period, the Navy distinguished between "line" and "engineering" officers. Ibid., p. 38.
39. Ibid., p. 25.
40. Ibid., p. 28.
41. Ibid., 31.
42. Ibid., p. 17.
43. Ibid., p. 222 (emphasis in original).
44. Allan Millett, "Assault from the Sea: The Development of Amphibious Warfare Between the Wars, the American, British, and Japanese Experiences," in Williamson Murray and Allan R. Millett, eds., *Military Innovation in the Interwar Period* (Cambridge: Cambridge University Press, 1996).
45. Ibid.
46. David E. Johnson, *Fast Tanks and Heavy Bombers: Innovation in the U.S. Army, 1917–1945* (Ithaca: Cornell University Press, 1998), p. 220.
47. Ibid., p. 226.
48. Emily O. Goldman and Leslie C. Eliason, eds., *The Diffusion of Military Technology and Ideas* (Stanford: Stanford University Press, 2003).
49. Emily O. Goldman, "Receptivity to Revolution: Carrier Air Power in Peace and War," in Goldman and Eliason, pp. 301–303.
50. Harvey M. Sapolsky, *The Polaris System Development: Bureaucratic and Programmatic Success in Government* (Cambridge: Harvard University Press, 1972), pp. 44, 59.

51. Ibid., p. 250.
52. Evangelista, p. 30.
53. Ibid.
54. Ibid., p. 42.
55. Tom Burns and G.M. Stalker, *The Management of Innovation* (London: Tavistock Publications, 1961), p. 119.
56. Ibid., pp. 120–121.
57. Balaji Chakravarthy, "Adaptation: A Promising Metaphor for Strategic Management," *Academy of Management Review*, Vol. 7, No. 1 (January 1982), p. 38.
58. Max Weber, *The Theory of Social and Economic Organization*, trans. A.M. Henderson and Talcott Parsons (New York: Free Press, 1964), p. 335.
59. Eric Heginbotham, "The British and American Armies in World War II: Explaining Variations in Organizational Learning Patterns," Defense and Arms Control Studies Program Working Paper No. 96–2, Massachusetts Institute of Technology, Cambridge, Mass., February 1996. Foley, "A Case Study in Horizontal Military Innovation: The German Army, 1916–1918," describes a similar process within the German Army during World War I.
60. Williamson Murray and Allan Millett, *Military Innovation in the Interwar Period*.
61. Williamson Murray, "Armored Warfare: The British, French, and German Experiences," in Williamson Murray and Allan Millett, *Military Innovation in the Interwar Period*.
62. Huntington, *The Soldier and the State*, p. 75.
63. Ibid., p. 76.
64. Posen, *The Sources of Military Doctrine*, p. 226.
65. Ibid., pp. 225–226, 236.
66. Sapolsky, *The Polaris System Development: Bureaucratic and Programmatic Success in Government*, p. 251.
67. See Andrew J. Bacevich, *The Pentomic Era: The U.S. Army Between Korea and Vietnam* (Washington, D.C.: National Defense University Press, 1986), pp. 19–70.
68. See Adam Grissom, "The Future of Military Innovation Studies," pp. 911–912.
69. Stephen Rosen's work points to a potential third model of successful adaptation, coming from within a military organization, from the bottom up. See Rosen, *Winning the Next War*, pp. 20–21. My prior research evaluated this model in the case studies and found it to be a "straw man." The wartime environment (at least in the case studies examined) precluded the opportunity for "new promotion pathways" to allow adaptations to percolate to the top of military organizations. Additionally, in most cases, new career paths existed prior to the wars.
70. Wesley Cohen and Daniel Levinthal, "Absorptive Capacity: A New Perspective on Learning and Innovation," *Administrative Sciences Quarterly*, Vol. 35, No.1 (March 1990), pp. 128–152.
71. Richard M. Cyert and James G. March, *A Behavioral Theory of the Firm* (Englewood Cliffs, N.J.: Prentice-Hall, 1963), p. 278.

72. Ibid., p. 279. This application of the business administration literature to military innovation is not without precedent. Matthew Evangelista argues that slack and knowledge base help determine how innovative a military organization is. The greater the slack, the more receptive the organization is to innovation. He also notes that organizational complexity, or the levels of skill and education of the organization's personnel, is "closely associated with innovation." Evangelista, *Innovation and the Arms Race*, pp. 45, 38.
73. Alexander L. George and Andrew Bennett, *Case Studies and Theory Development in the Social Sciences* (Cambridge: MIT Press, 2005), p. 6.
74. Having established "conceptual frameworks" and "recurring empirical regularities," I have built an explanatory model. This was used to rigorously collect "diagnostic evidence that provides the basis for casual inference." David Collier, "Understanding Process Tracing," *PS: Political Science and Politics*, Vol. 44, No. 4 (2011), p. 824. Establishing timelines of interest and gathering evidence of rival hypotheses was also vital. See also Jacob Ricks and Amy Liu, "Process Tracing: A Practical Guide," *PS: Political Science and Politics*, Vol. 51, No. 4 (October 2018), pp. 842–846.
75. Wayne P. Hughes, Jr., *Fleet Tactics and Coastal Combat*, 2nd ed. (Annapolis: Naval Institute Press, 2000), p. 7.
76. George and Bennett, *Case Studies and Theory Development in the Social Sciences*, pp. 178–179.

Chapter 2

1. Robert K. Massie, *Castles of Steel: Britain, Germany, and the Winning of the Great War at Sea* (New York: Random House, 2003), pp. 122–123.
2. Robert J. Art, "The Influence of Foreign Policy on Seapower: New Weapons and Weltpolitik in Wihelminian Germany," Sage Professional Paper in International Studies No. 02–019, Sage Publications, Beverly Hills, Calif., 1973, p. 13; and Bernard Brodie, *Sea Power in the Machine Age* (Princeton: Princeton University Press, 1944), p. 297.
3. Art, "The Influence of Foreign Policy on Seapower: New Weapons and Weltpolitik in Wihelminian Germany," pp. 28–33. For a description of the first "counterforce" (as Art describes it) U-boat cruise, see R.H. Gibson and Maurice Pendergast, *The German Submarine War, 1914–1918* (Great Britain: Constable and Company Ltd., 1931), pp. 2–3.
4. For the nature of the German U-boat campaign, see John Terraine, *The U-Boat Wars, 1916–1945* (New York: G.P. Putnam's Sons, 1989).
5. Geoffrey Bennett, *Naval Battles of the First World War* (London: Penguin Books, 2001), p. 265.
6. Ibid., p. 264.
7. Gibson and Pendergast, *The German Submarine War, 1914–1918*, pp. 32, 41.

8. Terraine, *The U-Boat Wars, 1916–1945*, p. 5.
9. Brodie, *Sea Power in the Machine Age*, p. 291.
10. Director of Trade Division, Admiralty, Memorandum, February 3, 1915, ADM 137, National Archives, Kew, England.
11. Commander in Chief, Grand Fleet Admiral Jellicoe, Undated Memorandum to Admiralty, ADM 137, National Archives, Kew, England.
12. Admiralty Memorandum, "Defense of Trade Against Submarine Attack," April 1, 1915, ADM 137, National Archives, Kew, England.
13. Bernard Ireland and Eric Grove, *Jane's War at Sea, 1897–1997, Centennial Edition* (New York: HarperCollins, 1997), p. 109.
14. See Table, Naval Staff, Admiralty, *The Technical History and Index, Volume 1, Part 7: The Anti-Submarine Division of the Naval Staff, December 1916–November 1918*, July 1919, Imperial War Museum, London, pp. 7–8.
15. See Viscount Jellicoe of Scapa, *The Crisis of the Naval War* (London: Cassel and Company Ltd., 1920).
16. Naval Staff, Training and Staff Duties Division, *Naval Staff Monograph (Historical), The Naval Staff of the Admiralty: Its Work and Development*, September 1929, BR1845, ADM 234/434, National Archives, Kew, England, p. 55.
17. Admiralty Memorandum and Responses, December 7–11, 1914, ADM 137, National Archives, Kew, England.
18. Ibid.
19. Ibid.
20. Admiral John Jellicoe, Commander in Chief Grand Fleet to Admiralty, Memorandum, December 15, 1914, ADM 137, National Archives, Kew, England.
21. Ibid.
22. Admiral John Jellicoe, Commander in Chief Grand Fleet to Admiralty, Memorandum, June 18, 1915, ADM 137, National Archives, Kew, England.
23. Bernard Brodie, *A Guide to Naval Strategy*, 5th ed. (New York: Praeger, 1965), p. 260.
24. Richard Hough, *The Great War at Sea* (Edinborough: Birlinn Ltd., 2000), p. 276.
25. Ibid., p. 306 (emphasis in original).
26. "Offensive measures" meant that, absent destroying U-boat bases, the BRN would address the problem more by hunting submarines, less by protecting merchant vessels.
27. Naval Staff, Training and Staff Duties Division, *Naval Staff Monograph (Historical), The Naval Staff of the Admiralty: Its Work and Development*, p. 62.
28. Ibid.
29. Ibid.
30. Naval Staff, Training and Staff Duties Division, *Naval Staff Monograph (Historical), The Naval Staff of the Admiralty: Its Work and Development*, p. 74. The Anti-Submarine Committee, formed in 1910, was a relatively small body charged with finding technological countermeasures to the submarine. It was not part of the BRN

staff. Chris Henry, *Depth Charge! Royal Naval Mines, Depth Charges, and Underwater Weapons, 1914–1945* (South Yorkshire: Pen and Sword Books Ltd., 2005), p. 21.
31. Memorandum by Director, Trade Division, Admiralty Staff, February 3, 1915, ADM 137, National Archives, Kew, England.
32. Memorandum by Director, Trade Division, Admiralty Staff, July 9, 1915, ADM 137, National Archives, Kew, England.
33. Memorandum by Director, Trade Division, Admiralty Staff, April 4, 1915, ADM 137, National Archives, Kew, England.
34. Admiralty Staff, "Considerations on Defense of Trade in the Channel and North Sea," Memorandum, April 1, 1915, ADM 137, National Archives, Kew, England.
35. The debate is documented in the following correspondence: Assistant Director of Operations, Admiralty Staff, "Response to 'Considerations on Defense of Trade,'" Memorandum, April 10, 1915; "Summary of Positions of Director of Trade Division and Director of Operations Division," Memorandum, April 11, 1915; Admiralty Staff, "Defense Against Submarines," Undated Memorandum, ADM 137, National Archives, Kew, England.
36. Director, Anti-Submarine Division Memorandum, February 19, 1917, ADM 137, National Archives, Kew, England.
37. Naval Staff, Admiralty, *The Technical History and Index, Volume 5, Part 7: Anti-Submarine Development: Experiments Prior to 1916*, September 1920, Imperial War Museum, London, p. 46.
38. Naval Staff, Admiralty, p. 48.
39. Commander, Third Battle Squadron to Commander in Chief, Grand Fleet, Memorandum, April 6, 1916, ADM 137, National Archives, Kew, England.
40. Naval Staff, Admiralty, *The Technical History and Index, Volume 1, Part 7*, p. 5. These officers also proved to be the most effective interrogators of captured U-boat crewmen, because they spoke the same technical language as their adversaries; H.D. Edwards to Submarine Depot Aberdeen, Memorandum, June 25, 1915, ADM 137, National Archives, Kew, England.
41. Admiralty Staff, "Report of Convoys," Memorandum, July 14, 1917, ADM 137, National Archives, Kew, England.
42. Director of Trade Division, Memorandum, December 2, 1918, ADM 137, National Archives, Kew, England.
43. *Jane's Fighting Ships for 1919*, Jane's Information Group, http://www.pbenyon1.plus.com/Janes_1919/Destroyers/Index.html, accessed April 24, 2009.
44. Gibson and Pendergast, *The German Submarine War, 1914–1918*, p. 149.
45. The British had emplaced 18,000 indicator buoys and 9,375 miles of wire for the nets by January 1915. Ibid., p. 50.
46. Robert Grant, *U-Boats Destroyed: The Effects of Anti-Submarine Warfare, 1914–1918* (London: Putnam, 1964), p. 161.
47. Memorandum, Admiral of Patrols to Admiralty, January 9, 1915, ADM 137, National Archives, Kew, England.

48. John Brooke, Personal Papers, Imperial War Museum, London, England.
49. Admiralty Memorandum, January 13, 1915, ADM 137, National Archives, Kew, England.
50. V.J. Robinson, Personal Papers, Imperial War Museum, London, England, pp. 50–60.
51. Admiralty Staff, "Defense of Trade Against Submarine Attack," April 1, 1915.
52. War Office to Admiralty Memorandum, December 28, 1915, and Admiralty to War Office Response, January 2, 1916, ADM 137, National Archives, Kew, England.
53. Senior Naval Officer Gibraltar to Admiralty, Memorandum, June 1, 1915 and Admiralty Response June 6, 1915, ADM 137, National Archives, Kew, England.
54. Vice Admiral Commanding Orkneys and Shetlands Memorandum, August 28, 1916, ADM 137, National Archives, Kew, England.
55. Naval Staff, Admiralty, *The Technical History and Index, Volume 1, Part 7: The Anti-Submarine Division of the Naval Staff, December 1916–November 1918*, July 1919, Imperial War Museum, London, p. 7.
56. Grant, *U-Boats Destroyed: The Effects of Anti-Submarine Warfare, 1914–1918*, p. 159.
57. Ibid., p. 161.
58. Brodie, *Sea Power in the Machine Age*, p. 305.
59. Gibson and Pendergast, *The German Submarine War, 1914–1918*, p. 52.
60. Ibid., p. 58.
61. Ibid., p. 79.
62. Ibid., p. 89.
63. Massie, *Castles of Steel: Britain, Germany, and the Winning of the Great War at Sea*, pp. 685, 690.
64. Naval Staff, Training and Staff Duties Division, p. 72.
65. Gibson and Pendergast, *The German Submarine War, 1914–1918*, p. 92.
66. Ibid., p. 93.
67. Naval Staff, Training and Staff Duties Division, p. 73.
68. Terraine, *The U-Boat Wars, 1916–1945*, p. 41.
69. Jellicoe, *The Crisis of the Naval War*, pp. 1–31.
70. W. Graham Green, Admiralty Office Memorandum, December 16, 1916, ADM 137, National Archives, Kew, England.
71. Massie, *Castles of Steel: Britain, Germany, and the Winning of the Great War at Sea*, p. 736.
72. See Henry, *Depth Charge! Royal Naval Mines, Depth Charges, and Underwater Weapons, 1914–1945*.
73. These calculations are based on Grant, *U-Boats Destroyed: The Effects of Anti-Submarine Warfare, 1914–1918*, p. 159.
74. Terraine, *The U-Boat Wars, 1916–1945*, p. 50.
75. Hough, *The Great War at Sea*, p. 307.

76. Admiralty Staff, Minutes of Meeting, February 23, 1917, ADM 137, National Archives, Kew, England.
77. See, for example, Massie, *Castles of Steel: Britain, Germany, and the Winning of the Great War at Sea*, p. 729.
78. Captain H. Kerwick, Royal Navy Reserve, Admiralty Staff, Memorandum, February 10, 1919, ADM 137, National Archives, Kew, England.
79. Paul G. Halpern, *A Naval History of World War I* (Annapolis: Naval Institute Press, 1994), p. 355.
80. This euphemism is explained in the discussion of the French coal trade below. Naval Staff, Training and Staff Duties Division, p. 75.
81. Halpern, *A Naval History of World War I*, p. 352.
82. Naval Staff, Training and Staff Duties Division, p. 75.
83. Halpern, *A Naval History of World War I*, p. 353.
84. Ibid., p. 73.
85. Central Executive Committee of Great Britain to Admiralty, Letter, December 28, 1916, ADM 137, National Archives, Kew, England.
86. When an important document was circulated through the Admiralty Staff, each division representative made their comments on a routing sheet, much like in any other bureaucratic organization. The word "endorsement" does not imply advocacy; instead, it indicates that the actor in question has read the document in question and added his comments to it.
87. Central Executive Committee of Great Britain Letter, Trade Division Endorsement, January 1, 1917.
88. Ibid., Anti-Submarine Division Endorsement.
89. Ministry of Marine, Paris, to Admiralty, Telegram, Sent 2155, December 28, 1916, ADM 137, National Archives, Kew, England.
90. Admiralty to Central Executive Committee of Great Britain, Letter, January 16, 1917, ADM 137, National Archives, Kew, England.
91. Operations Division, Admiralty Staff, Memorandum, "Mission of Commandant Vandier," January 3, 1917, ADM 137, National Archives, Kew, England.
92. Ibid.
93. Ibid.
94. Operations Division Endorsement, "Mission of Commandant Vandier," January 3, 1917.
95. Naval Staff, Training and Staff Duties Division, p. 76 and note.
96. Operations and Trade Divisions Endorsements, "Mission of Commandant Vandier," January 3, 1917.
97. Director, Antisubmarine Division Endorsement, "Mission of Commandant Vandier," January 4, 1917.
98. Anthony Downs, *Inside Bureaucracy* (Prospect Heights, Illinois: Waveland Press, 1994), p. 84. One must be careful here about appearing to make a normative judgment about "personal effort." My research indicates that the Admiralty Staff was

indeed overworked. It is doubtful that "laziness" or the like had anything to do with this instance of bureaucratic politics; rather, it was the incapacity of these subunits to objectively handle an increase in workload.

99. Naval Staff, Training and Staff Duties Division, p. 76.

100. Admiralty Memorandum Endorsement, "Mission of Commandant Vandier," January 16, 1917.

101. Massie, *Castles of Steel: Britain, Germany, and the Winning of the Great War at Sea*, p. 730.

102. Halpern, *A Naval History of World War I*, p. 360.

103. Ibid., p. 361.

104. H.W. Kenrick, Royal Navy Reserve, "History of the Admiralty Shipping Intelligence Office—London," January 1919, ADM 137, National Archives, Kew, England.

105. Admiralty Staff Training Division, p. 79.

106. Safety of Transports, Admiralty Memorandum, February 2, 1915, ADM 137, National Archives, Kew, England.

107. Admiralty, Notes of a Meeting to Consider the Convoy System, December 14, 1917, ADM 137, National Archives, Kew, England.

108. Admiralty Staff, "The Convoy System," Memorandum Appendix I, September 17, 1917, ADM 116/1349, National Archives, Kew, England.

109. Admiralty Staff, "The Policy of Defense of Trade," Memorandum, September 17, 1917, ADM 116, National Archives, Kew, England.

110. Admiralty Staff, Memorandum on the Mercantile Convoy System, September 3, 1917, ADM 116, National Archives, Kew, England.

111. Commander in Chief Plymouth to Admiralty, Memorandum, November 5, 1917, ADM 137, National Archives, Kew, England.

112. Commander in Chief Plymouth to Admiralty, Memorandum, August 7, 1918; Director, Mercantile Movements Division Note, February 23, 1918, ADM 137, National Archives, Kew, England.

113. Admiralty Staff, "History of Shipping Intelligence Office," January 1919, ADM 137, National Archives, Kew, England.

114. Naval Staff, Admiralty, *The Technical History and Index, Volume 1, Part 7: The Anti-Submarine Division of the Naval Staff, December 1916–November 1918*, July 1919, Imperial War Museum, London, pp. 7–8.

115. See Naval Staff, Training and Staff Duties Division, *Naval Staff Monograph (Historical), The Naval Staff of the Admiralty: Its Work and Development*, September 1929, BR1845, ADM 234/434, National Archives, Kew, England.

116. Naval Staff, Training and Staff Duties Division, p. 53.

117. Global Security.org, "First Sea Lord—1889–1919," http://www.globalsecurity.org/military/world/europe/uk-rn-fsl2.htm, accessed October 28, 2008.

118. Naval Staff, Training and Staff Duties Division, p. 73.

119. Naval Staff, Training and Staff Duties Division, p. 75.

120. Terraine, *The U-Boat Wars, 1916–1945*, pp. 53–54; and Halpern, *A Naval History of World War I*, p. 355.
121. Massie, *Castles of Steel: Britain, Germany, and the Winning of the Great War at Sea*, p. 122.
122. See Brodie, *Sea Power in the Machine Age*, pp. 300–302.
123. Philip J. Haythornthwaite, *The World War One Source Book* (London: Arms and Armour Press, 1992), p. 105.
124. Admiralty Memorandum, November 26, 1917, ADM 116, National Archives, Kew, England.
125. Haythornthwaite, *The World War One Source Book*, pp. 106–107.
126. Art, "The Influence of Foreign Policy on Seapower: New Weapons and Weltpolitik in Wihelminian Germany," p. 15.
127. Admiral John J. Jellicoe, "The Influence of the Submarine Upon Naval Policy and Operations," First Sea Lord Memorandum to the War Cabinet, November 18, 1917, ADM 116, National Archives, Kew, England.
128. Ibid.
129. Admiralty Memorandum, November 26, 1917, ADM 116, National Archives, Kew, England.
130. Admiral Beatty, Commander In Chief, Grand Fleet to Admiralty, Memorandum, January 9, 1918, ADM 116, National Archives, Kew, England.
131. First Lord of the Admiralty to War Cabinet, Memorandum, January 17, 1918, ADM 116, National Archives, Kew, England.
132. Naval Staff, Admiralty, *The Technical History and Index, Volume 1, Part 7*, p. 7.
133. Admiral John Jellicoe, Commander In Chief, Grand Fleet to Admiralty, Memorandum, December 11, 1914, ADM 137, National Archives, Kew, England.
134. Admiralty War Staff, Trade Division, "Instructions for Owners and Masters of British Merchant Ships Issued With Reference to the Operations of German Submarines against British Shipping," Pamphlet, February 10, 1915, ADM 137, National Archives, Kew, England.
135. Admiralty War Staff, Trade Division, "Trade Routes During War," Pamphlet, August 4, 1915, ADM 137, National Archives, Kew, England.
136. Commander In Chief, Grand Fleet (HMS *Iron Duke*) Memorandums February 1 and 14, 1915, ADM 137, National Archives, Kew, England.
137. Director of Naval Operations, Naval War Staff, Memorandum, February 2, 1915, ADM 137, National Archives, Kew, England (emphasis in original).
138. "Policy in Defense of Trade," Admiralty Memorandum Enclosure, January 1918, ADM 116, National Archives, Kew, England.
139. Numbers rounded to nearest 1000 tons and calculated from Table, Naval Staff, Admiralty, *The Technical History and Index, Volume 1, Part 7: The Anti-Submarine Division of the Naval Staff, December 1916–November 1918*, July 1919, Imperial War Museum, London, p. 7.

140. Massie, *Castles of Steel: Britain, Germany, and the Winning of the Great War at Sea*, p. 730.
141. See Jan S. Breemer, "Defeating the U-boat: Inventing Antisubmarine Warfare," Newport Paper No. 36, Naval War College Press, Newport, R.I., August 2010, pp. 58–61.
142. Breemer notes that "the evidence is mostly circumstantial, but it is difficult to avoid the conclusion that" Lloyd George's account carries more weight. Ibid.
143. Ibid., p. 60.
144. Hough, *The Great War at Sea*, p. 309.
145. Quoted in Terraine, *The U-Boat Wars, 1916–1945*, p. 59. See also Edwyn Gray, *The U-boat War* (London: Leo Cooper, 1994), p. 182.
146. Massie, *Castles of Steel: Britain, Germany, and the Winning of the Great War at Sea*, p. 730.
147. Submarine Attack Committee Minutes, Anti-Submarine Menace, August 17, 1915, ADM 137, National Archives, Kew, England. This document also notes that the British Royal Navy took up as many trawlers as possible (without overly disrupting fishing) by this point in the war.
148. Grand Fleet Battle Orders, August 31, 1914, ADM 137, National Archives, Kew, England.
149. The Destroyer Position in Home Waters, Admiralty Memorandum, undated, ADM 137, National Archives, Kew, England.
150. Gibson and Pendergast, *The German Submarine War, 1914–1918*, p. 153.
151. See, for example, Dwight R. Messimer, *Find and Destroy: Antisubmarine Warfare in World War One* (Annapolis: Naval Institute Press, 2001), p. 147.
152. Terraine, *The U-Boat Wars, 1916–1945*, p. 39.
153. Art, "The Influence of Foreign Policy on Seapower: New Weapons and Weltpolitik in Wihelminian Germany," p. 13.

Chapter 3

1. Atsushi Oi, "The Japanese Navy in 1941," in Donald M. Goldstein and Katherine V. Dillon, eds., *The Pacific War Papers: Japanese Documents of World War II* (Washington, D.C.: Potomac Books, 2004), p. 12.
2. While tactically indecisive, the 1916 Battle of Jutland between the British Royal and Imperial German Navies was one of the few instances of engagements between the principal battle forces of two navies. Alfred Thayer Mahan argued that the destruction of the enemy's fleet was necessary for "command of the sea," which in turn enabled victory in war. Other operations, for example commerce raiding, were indecisive to the ultimate outcome of a war and therefore a misuse of the fleet. Adherents of Mahan's theory sought, as the logical conclusion of Mahan's argument, the "decisive engagement" with the enemy's main fleet as the means to gain command of the sea.

"Jutland" and "Mahanian decisive battle" are thus metaphors for a battle between the main strengths of two fleets that could lead to a decisive outcome in the struggle for control of the sea in a conflict. Nathan Miller, *War At Sea: A Naval History of World War II* (New York: Scribner, 1995), pp. 303–310; Russel F. Weigley, *The American Way of War: A History of United States Military Strategy and Policy* (Bloomington: Indiana University Press, 1973), pp. 245–246. See also Edward Miller, "Kimmel's Hidden Agenda," *MHQ: The Quarterly Journal of Military History*, Vol. 4, No. 1 (Autumn 1991), pp. 36–43, and Edward Miller, *War Plan Orange: The U.S. Strategy to Defeat Japan, 1897–1945* (Annapolis: Naval Institute Press, 1991), pp. 303–310.

3. Oi, "The Japanese Navy in 1941," p. 20.
4. Ibid., p. 19.
5. Mark R. Peattie, *Sunburst: The Rise of Japanese Naval Air Power, 1909–1941* (Annapolis: Naval Institute Press, 2001), p. 128.
6. Ibid., p. 129.
7. Weigley, *The American Way of War: A History of United States Military Strategy and Policy*, p. 265.
8. Peattie, *Sunburst: The Rise of Japanese Naval Air Power, 1909–1941*, p. 140.
9. Ibid., p. 141.
10. Ibid., p. 144.
11. Ibid., p. 140.
12. Ibid., p. 143.
13. Ibid., pp. 149–151.
14. Gordon W. Prange, with Donald M. Goldstein and Katherine V. Dillon, *At Dawn We Slept: The Untold Story of Pearl Harbor* (New York: McGraw Hill, 1981), pp. 101–102, 106; Peattie, *Sunburst: The Rise of Japanese Naval Air Power, 1909–1941*, p. 152.
15. Alfred T. Mahan, *The Influence of Sea Power Upon History, 1660–1783* (New York: Dover Publications, 1987).
16. Peattie, *Sunburst: The Rise of Japanese Naval Air Power, 1909–1941*, p. 170.
17. Ibid., p. 131.
18. Ibid.
19. Ibid.
20. Masatika Chihaya, "Some Features Concerning Changes in the Japanese Fleet Organization During the War," Goldstein and Dillon, p. 285.
21. Peattie, *Sunburst: The Rise of Japanese Naval Air Power, 1909–1941*, p. 134.
22. Oi, "The Japanese Navy in 1941," p. 24.
23. Ibid., p. 25.
24. Lisle A. Rose, *Power at Sea, Volume 2: The Breaking Storm, 1919–1945* (Columbia: University of Missouri Press, 2007), p. 145.
25. Chihaya, "Some Features Concerning Changes in the Japanese Fleet Organization During the War," p. 286.
26. Peattie, *Sunburst: The Rise of Japanese Naval Air Power, 1909–1941*, p. 157.

27. Ibid., p. 156.
28. Ibid., pp. 158–159.
29. Ibid., p. 156.
30. Ibid., p. 158
31. Rose, *Power at Sea, Volume 2: The Breaking Storm, 1919–1945*, p. 260.
32. For this view see, for example, Patrick Degan, *Flattop Fighting in World War II: The Battles Between American and Japanese Aircraft Carriers* (Jefferson, N.C.: MacFarland and Company, 2003), p. 52.
33. Staff Presentation, "Employment of Battleships," Naval War College, June 9, 1943, Naval War College Historical Collection, Record Group 14, p. 21.
34. U.S. Navy, "FTP 143(A): War Instructions, United States Navy, 1944," under "Doctrine and Tactics Publications," www.ibiblio.org.hyperwar/USN/ref/index.html, accessed February 3, 2009.
35. George C. Dyer, *On the Treadmill to Pearl Harbor: The Memoirs of Admiral J.O. Richardson* (Washington, D.C.: Navy History Division, Department of the Navy, 1973), p. 220.
36. Ibid., p. 222.
37. Commander in Chief, Pacific Fleet, "Annual Report of the Commander in Chief, U.S. Pacific Fleet, for the Period 1 July 1940 to 30 June 1941," Records of the U.S. Navy General Board, National Archives, Washington, D.C.
38. Office of the Chief of Naval Operations, *United States Naval Aviation in the Pacific*, Department of the Navy 1947, Naval Aviation History Office, www.history.navy.mil/branches/org4-23, accessed February 3, 2009, p. 2.
39. Julius Augustus Furer, *Administration of the Navy Department in World War II* (transcribed and formatted for HTML by Patrick Clancey, HyperWar Foundation), www.ibiblio.org/hyperwar/USN/-Admin/index.html, accessed August 12, 2009, p. 365.
40. Bureau of Naval Personnel, Training, Standards and Curriculum Division, *Seamanship*, NAVPERS 16118, June 1944, www.ibiblio.org/hyperwar/USN/ref/Seamanship/, accessed August 12, 2008, pp. 18–21.
41. Office of War Information, Division of Public Inquiries, "Department of the Navy," in *United States Government Manual*, 1st ed., 1945, www.ibiblio.org/hyperwar/ATO/USGM/Navy.html, p. 310.
42. Department of the Navy, Naval Historical Division, *U.S. Naval Administrative Histories of World War II Number 27, DCNO(Air): Aviation in the Office of the CNO—the establishment of DCNO(Air)*, Naval Historical Center Archives, Washington Navy Yard, p. 6.
43. Ibid., p. 11.
44. Ibid., p. 18.
45. Ibid., p. 16.
46. Ibid., p. 304.
47. Furer, *Administration of the Navy Department in World War II*, p. 364.

48. Bureau of Naval Personnel, Training, Standards and Curriculum Division, p. 18. Gates's influence on naval matters is evident through the fact that an entire record group at the National Archives in College Park, Maryland, is reserved for his official correspondence.
49. See Emily O. Goldman, "Receptivity to Revolution: Carrier Air Power in Peace and War," in *The Diffusion of Military Technology and Ideas*, Emily Goldman and Leslie C. Eliason, eds., (Stanford: Stanford University Press, 2003), p. 286; and Geoffrey Till, "Adopting the Aircraft Carrier: the British, American, and Japanese Case Studies," in Williamson Murray and Allan R. Millett, eds., *Military Innovation in the Interwar Period* (Cambridge: Cambridge University Press, 1996), p. 211.
50. Thomas B. Buell, *Master of Sea Power: A Biography of Fleet Admiral Ernest J. King* (Annapolis: Naval Institute Press, 1980), pp. 72, 79.
51. Ibid., pp. 80–93.
52. Rear Admiral Francis D. Foley, USN (ret.), interviewed by Paul Stillwell, January 16, 1985, Interview Number 3, transcript, Record Group OH 120, Naval War College Historical Collection, Newport, Rhode Island, p. 225.
53. Buell, p. 153.
54. Ibid., p. 179 (emphasis in original).
55. Ibud., p. 153.
56. Admiral Ernest J. King, *Our Navy at War: A Report to the Secretary of the Navy*, March 27, 1944, Record Group 4, Naval War College Historical Collection, Newport, Rhode Island, p. 25.
57. Ibid.
58. Ibid.
59. Miller, p. 279.
60. Marc Bernstein, "The Early Carrier Raids: Proving Japanese Vulnerability," in *Naval Aviation News*, Naval Aviation History Office, March–April 1992, www.history.navy.mil/branches/article-2.html, accessed February 3, 2009, p. 28.
61. John Elliot, "Aviation Ordnance, 1939–1941," in *Naval Aviation News*, Naval Aviation History Office, July–August 1991, www.history.navy.mil/branches/article-2.html, accessed February 3, 2009, p. 27.
62. Fighter direction was the use of shipboard radar and personnel to coordinate interception of enemy attack aircraft before they reached the task force. See Weigley, *The American Way of War: A History of United States Military Strategy and Policy*, p. 271.
63. Ibid.
64. Samuel Eliot Morison, *History of United States Naval Operations in World War II*, Vol. 4, *Coral Sea and Midway* (Chicago: University of Illinois Press, 1960), pp. 142–143; and Thomas B. Buell, *The Quiet Warrior: A Biography of Admiral Raymond A. Spruance* (Boston: Little, Brown and Company, 1974), p. 140.
65. Morison, *History of United States Naval Operations in World War II*, Vol. 4, *Coral Sea and Midway*, p. 149; Buell, *The Quiet Warrior*, pp. 145–147.

66. Commander in Chief, Pacific Fleet to Commander in Chief, U.S. Fleet, undated memorandum, "Battle of Midway," Bureau of Aeronautics Confidential Correspondence, Record Group 72, National Archives, College Park, Maryland, p. 27.
67. Commander in Chief, U.S. Fleet, "Action Reports of Yorktown Air Group During Coral Sea Engagement," undated memorandum, Bureau of Aeronautics Confidential Correspondence, Record Group 72, National Archives, College Park, Maryland.
68. Commander J.M. Worthington, U.S. Navy, Staff Presentation, "Battle of Midway, 3 to 6 June 1942," Naval War College, August 28, 1943, Naval War College Historical Collection, Record Group 14, Newport, Rhode Island, p. 24; Letter from Raymond Spruance to Chester Nimitz, June 8, 1942; Commander, Task Force 16 to Commander in Chief, U.S. Pacific Fleet, memorandum, "Battle of Midway, Forwarding of Reports," June 16, 1942, Raymond O. Spruance Papers, Naval War College Historical Collection, Newport, Rhode Island.
69. Buell, *The Quiet Warrior*, p. 153.
70. Commander in Chief, Pacific Fleet, "Battle of Midway," p. 25.
71. Commander in Chief, U.S. Fleet, "Action Reports of Yorktown Air Group During Coral Sea Engagement."
72. "Battle of Midway, 3 to 6 June 1942," p. 25.
73. Ibid.
74. Captain R.W. Bates, U.S. Navy, Staff Presentation, "Anti-Aircraft Defense," Naval War College, June 17, 1943, Naval War College Historical Collection, Record Group 14, Newport, Rhode Island, pp. 10–11.
75. Ibid., p. 18.
76. Commander in Chief, U.S. Pacific Fleet to Chairman, General Board, memorandum, "Are We Ready III," Records of the U.S. Navy General Board, National Archives, Washington, D.C., p. 55.
77. Commander in Chief, U.S. Pacific Fleet, "Are We Ready III," enclosure (A), p. 28.
78. See, for example, Commander, Carrier Replacement Group Number Ten to Commander, Carriers Pacific Fleet, memorandum, "Readiness for war—report of," July 1, 1942, Bureau of Aeronautics Confidential Correspondence, Record Group 72, National Archives, College Park, Maryland. Replacement Air Groups were the conduits through which aviators passed on their way from basic flight training to front line squadrons.
79. *U.S. Naval Aviation, 1910–1945, Part 5: World War II, 1940–1945*, p. 111.
80. Office of the Chief of Naval Operations, p. 11.
81. Kenneth N. Waltz, *Theory of International Politics* (Boston: McGraw-Hill, 1979), p. 105.
82. Department of the Navy, Naval Historical Division, *U.S. Naval Administrative Histories of World War II Number 89: An Administrative History of the Bureau of Ships in World War II*, Vol. I, Naval Historical Center Archives, Washington Navy Yard, p. 17.
83. Ibid., p. 94.

84. Ibid., p. 117.
85. *U.S. Naval Administrative Histories of World War II Number 89*, Vol. II, p. 207.
86. Buell, *Master of Sea Power*, pp. 308–311.
87. *U.S. Naval Administrative Histories of World War II Number 89*, Vol. II, p. 333.
88. Ibid., Table 40.
89. Buell, *Master of Sea Power*, p. 356.
90. Miller, pp. 276–278.
91. Headquarters, Commander in Chief, U.S. Fleet, "Secret Information Bulletin Number 2: Battle Experience Solomon Islands Actions, August–September 1942," Naval War College Historical Collection, Newport, Rhode Island, pp. 12–15.
92. Ibid., pp. 12–14.
93. Ibid., pp. 12–16.
94. Ibid., pp. 282–284.
95. Commander in Chief, Pacific Fleet to Commander in Chief, U.S. Fleet, memorandum, "Solomon Islands Campaign—Action of 23–25 August," October 24, 1942, Bureau of Aeronautics Secret Correspondence, Record Group 72, National Archives, College Park, Maryland, p. 19.
96. R.L. Campbell, Staff Presentation, "The Anti-Aircraft Defense of a Task Force," Naval War College, November 9, 1944, Naval War College Historical Collection, Record Group 14, Newport, Rhode Island; and Commander, USS *North Carolina*, to Commander-in-Chief, U.S. Fleet (Readiness Division), memorandum, "Action of August 24, 1942," Bureau of Aeronautics Confidential Correspondence, Record Group 72, National Archives, College Park, Maryland.
97. "Secret Information Bulletin Number 2," pp. 12–13.
98. Rose, *Power at Sea, Volume 2: The Breaking Storm, 1919–1945*, p. 253.
99. "Secret Information Bulletin Number 2," pp. 12–19.
100. Commander in Chief, Pacific Fleet, "Solomon Islands Campaign—Action of 23–25 August," p. 17
101. "Secret Information Bulletin Number 2," pp. 12–19.
102. Foley interview, p. 384.
103. Foley Interview, p. 390.
104. "Secret Information Bulletin Number 2," pp. 12–15.
105. Bates, p. 29.
106. Hanson W. Baldwin, interview by John T. Mason, Jr., April 21, 1975, Interview Number 3, transcript, Record Group OH 72, Naval War College Historical Collection.
107. "Air Defense of a Task Force," p. 27.
108. The VT fuse was test-fired in January 1942 and first used in combat on January 5, 1943. *U.S. Naval Aviation, 1910–1945, Part 5: World War II, 1940–1945*, Naval Aviation History Office, www.history.navy.mil/branches/org4-23.htm, accessed February 3, 2009, p. 122.
109. "Secret Information Bulletin Number 2," pp. 12–28.

110. Peattie, *Sunburst: The Rise of Japanese Naval Air Power, 1909–1941*, p. 186; Jisaburo Ozawa, "Outline of Tactics and Organization of the Japanese Carrier Air Force," in Goldstein and Dillon, p. 80.
111. Weigley, *The American Way of War: A History of United States Military Strategy and Policy*, pp. 292–299; Buell, pp. 257–280; Miller, pp. 438–446; and James D. Hornfischer, *The Fleet at Flood Tide: America at Total War in the Pacific, 1944–1945* (New York: Bantam Books, 2016), pp. 178–203.
112. Worrell Reed Carder, Rear Admiral, U.S. Navy, *Beans, Bullets, Black Oil*, www.ibiblio.org/hyperwar/USN/-Admin/index.html, p. 4.
113. Thomas Wildenberg, *Grey Steel and Black Oil*, www.ibiblio.org/hyperwar/USN/-Admin/index.html, p. 28.
114. Carder, *Beans, Bullets, Black Oil*, p. 4.
115. Wildenberg, *Grey Steel and Black Oil*, p. 28.
116. Ibid., p. 37.
117. Ibid., p. 30.
118. Ibid., p. 128.
119. Ibid., p. 129.
120. Carder, *Beans, Bullets, Black Oil*, p. 11.
121. Ibid., p. 5.
122. Weigley, *The American Way of War: A History of United States Military Strategy and Policy*, p. 284.
123. Carder, *Beans, Bullets, Black Oil*, p. 5.
124. Mitsuo Fuchida and Masatake Okumiya, *Midway: The Battle That Doomed Japan* (Annapolis: Naval Institute Press, 1955), p. 208, quoted in Peattie, *Sunburst: The Rise of Japanese Naval Air Power, 1909–1941*, p. 192.
125. War Department General Staff, War Plans Division to Chief of Naval Operations, memorandum, "Strategic Estimate: Exhibit 3: Data Pertaining to Sea Forces," November 13, 1941, Bureau of Aeronautics Secret Correspondence, Record Group 72, National Archives, College Park, Maryland.
126. John A. Adams, *If Mahan Ran the Pacific War: An Analysis of World War II Naval Strategy* (Bloomington: Indiana University Press, 2008), p. 382.
127. Max Hastings, *Retribution: The Battle for Japan, 1944–45* (New York: Vintage Books, 2007), p. 393.
128. Military Historical Section, Headquarters, Army Forces Far East, "Outline of Naval Armament and Preparations for War," Japanese Monograph Number 174, Record Group 550, National Archives, College Park, Maryland, p. 4. This monograph was written from the Imperial Japanese perspective.
129. Ibid., p. 2.
130. "The Great Marianas Turkey Shoot," otherwise known as the Battle of the Philippine Sea in June 1944, saw Admiral Spruance's Fifth Fleet aircraft shoot down nearly 600 Japanese aircraft for a loss of only 123. Admiral King noted, "The great loss of Japanese naval aircraft in the Battle of the Philippine Sea crippled Japanese

naval aviation for the remainder of the war." James D. Hornfischer, *The Fleet at Flood Tide: America at Total War in the Pacific, 1944–1945* (New York: Bantam Books, 2016), pp. 224, 306; and Naval History and Heritage Command, "'The Great Marianas Turkey Shoot,' 19–20 June 1944," www.history.navy.mil/browse-by-topic/wars-conflicts-and-operations/world-war-ii/1944/battle-philippine-sea/turkey-shoot.html, accessed August 26, 2022.

131. Military Historical Section, Headquarters, Army Forces Far East, p. 9.

132. Ibid.

133. Vice Admiral Lloyd M. Mustin, USN (ret.), interviewed by John T. Mason, Jr., March 10, 1970, Transcript, Naval War College Historical Collection, p. 16.

134. Admiral John S. Thatch, USN (ret.), interviewed by Etta Belle Kitchen, January 9, 1971, Interview Number 4, transcript, Record Group OH 58, Naval War College Historical Collection, p. 389.

135. Miller, p. 474.

136. Adams, *If Mahan Ran the Pacific War: An Analysis of World War II*, p. 381.

137. Miller, p. 497.

138. Miller, p. 524.

139. Commanding Officer, USS *North Carolina* to Commander in Chief, U.S. Fleet, Readiness Division, "Action of August 24, 1942," Bureau of Aeronautics Confidential Correspondence, Record Group 72, National Archives, College Park, Maryland.

140. Mustin interview, p. 16.

141. Miller, p. 493.

142. Ibid., p. 492.

143. Ibid., p. 494.

144. Ibid., p. 491.

145. Ibid., p. 498.

146. Thatch interview, pp. 391–392.

147. Thatch interview, p. 394.

148. Miller, p. 498.

149. Ibid., pp. 513–514.

150. Rear Admiral Richard R. Ballinger, USN (ret.), interviewed by Barrie Grenell, December 1, 1976, Interview Number 3, transcript, Record Group OH 83, Naval War College Historical Collection, p. 40.

151. Naval War College Staff presentation, "Operations at Okinawa," November 17, 1945, Naval War College Historical Collection, Record Group 14, Newport, Rhode Island, p. 17.

152. Thatch interview, p. 390; "Operations at Okinawa," p. 17.

153. Miller, p. 521.

154. "Operations at Okinawa," p. 19. See also Mark Stille, *U.S. Navy Ships vs. Kamikazes, 1944–45* (New York: Osprey Publishing Ltd., 2016), pp. 37, 63–71.

155. Ibid., p. 18; Miller, p. 521.

156. "Operations at Okinawa," p. 18.

157. Ibid., p. 19.
158. Miller, p. 525.
159. If one assumes that "seriously damaged" caused at least a temporary loss to the picket line, that means that 55 percent of the pickets were put out of action. "Operations at Okinawa," p. 20.
160. Miller, p. 524.
161. "Operations at Okinawa," p. 19.
162. "Operations at Okinawa," p. 18; and Admiral Harry D. Felt, USN (ret.), interviewed by John T. Mason, Jr., March 6, 1972, Interview Number 3, transcript, Record Group OH 137, Naval War College Historical Collection, p. 166.
163. "Operations at Okinawa," p. 23.
164. Mustin interview, p. 14.
165. Ibid., p. 34.
166. Ibid., p. 22.
167. Ibid., p. 21.
168. Ibid., p. 34.
169. Military Historical Section, Headquarters, Army Forces Far East, p. 19
170. Miller, p. 526.

Chapter 4

1. Charles T. Kamps, "The JCS 94-Target List: A Vietnam myth that still distorts military thought," *Aerospace Power Journal*, Spring 2001, www.airpower.maxwell.af.mil/airchronicles/apj/apj01/spr01/kamps.pdf, p. 71; and Major Ricky J. Drake, USAF, "The Rules of Defeat: The impact of aerial rules of engagement on USAF operations in North Vietnam, 1965–1968," Master's Thesis, School of Advanced Airpower Studies, Air University, Maxwell Air Force Base, May 1992, p. 30.
2. Jacob Van Staaveren, "The Air Force in Southeast Asia: toward a bombing halt, 1968," Office of Air Force History, Maxwell Air Force Base, September 1970, p. 6.
3. Earl H. Tilford, Jr., *Crosswinds: The Air Force's Setup in Vietnam* (College Station: Texas A&M University Press, 1993), 81 (emphasis in original).
4. Ibid, p. 82.
5. Jacob Van Staaveren, "USAF Plans and Operations in Southeast Asia, 1965," USAF Historical Division Liaison Office, USAF Historical Branch Archives, Maxwell Air Force Base, October 1966, p. 72.
6. USAF Historical Division, Liaison Office, "The Air Campaign Against North Vietnam, 1966," p. 64.
7. The following discussion is based upon two sources: Anthony M. Thornborough and Frank B. Mormillo, et. al., *Iron Hand: Smashing the Enemy's Air Defenses* (Sparkford, England: Patrick Stephens Ltd., 2002), pp. 3–5; and USAF, "Tactics and Techniques of Electronic Warfare," Air Force Historical Branch, pp. 1–13.

8. Ibid., p. 8.
9. Ibid., p. 9.
10. Ibid.
11. Ibid., p. 1.
12. Thornbourough and Mormillo, *Iron Hand: Smashing the Enemy's Air Defenses*, p. 39.
13. For a detailed account on the rise and fall of the bomber generals, see Michael R. Worden, *Rise of the Fighter Generals: The Problem of Air Force Leadership. 1945–1982* (Maxwell Air Force Base, Alabama: Air University Press, 1998).
14. The SAC-TAC split and TAC inferiority within the USAF is well known. See, for example, Worden, *Rise of the Fighter Generals: The Problem of Air Force Leadership. 1945–1982*, Chapters 4–6; Mark Clodfelter, *The Limits of Air Power: The American Bombing of North Vietnam* (New York: Free Press, 1989), pp. 26–37; "TAC was just about out of business" before Vietnam, General William C. Norris, U.S. Air Force Oral History Interview transcript, Office of Air Force History, November 30, 1984, p. 45.
15. Craig C. Hannah, *Striving for Air Superiority: The Tactical Air Command in Vietnam* (College Station: Texas A&M University Press, 2002), pp. 47–48.
16. Norris, U.S. Air Force Oral History Interview transcript, p. 18.
17. Hannah, *Striving for Air Superiority: The Tactical Air Command in Vietnam*, p. 50.
18. Ibid., p. 51.
19. Ibid., p. 53.
20. Robert A. Pape, *Bombing to Win: Air Power and Coercion in War* (Ithaca: Cornell University Press, 1996), p. 177.
21. Ibid., pp. 178–179.
22. Ibid., pp. 180–181.
23. Ibid., p. 181.
24. Ibid., p. 184.
25. See Kenneth Bell, *100 Missions North* (New York: Brassey's [U.S.], 1993).
26. Pape, *Bombing to Win: Air Power and Coercion in War*, p. 188.
27. Van Staaveren, p. 37. The JCS considered mining critical to shutting down the port. See Kamps, "The JCS 94-Target List."
28. James R. Brungess, LTCOL USAF, *Setting the Context: suppression of enemy air defenses and joint war fighting in an uncertain world* (Maxwell Air Force Base, Alabama: Air University Press, 1994), pp. 5–9.
29. USAF, "Tactics and Techniques of Electronic Warfare," p. 32.
30. Thornbourough and Mormillo, *Iron Hand: Smashing the Enemy's Air Defenses*, p. 39.
31. USAF, "Tactics and Techniques of Electronic Warfare," p. 32.
32. Thornbourough and Mormillo, *Iron Hand: Smashing the Enemy's Air Defenses*, p. 40; Brungess, *Setting the Context: suppression of enemy air defenses and joint war fighting in an uncertain world*, p. 5.

33. Alfred Price, *War in the Fourth Dimension: US Electronic Warfare from the Vietnam War to the Present* (London: Greenhill Books, 2001), p. 44.
34. "Respectively, the letters denote piloted aircraft, radar, and receiving passive-detecting, i.e. installation class, equipment, and purpose." David R. Kendziora, "F-4G Wild Weasel," Office of History, U.S. Air Force, Ogden Air Logistics Center, Hill Air Force Base Utah, 1983, Air Force Historical Research Center, Maxwell Air Force Base, p. 1.
35. Thornbourough and Mormillo, *Iron Hand: Smashing the Enemy's Air Defenses*, p. 41.
36. Price, *War in the Fourth Dimension: US Electronic Warfare from the Vietnam War to the Present*, p. 75.
37. Ibid., p. 40.
38. Kendziora, "F-4G Wild Weasel," p. 2.
39. Thornbourough and Mormillo, *Iron Hand: Smashing the Enemy's Air Defenses*, p. 42.
40. Kendziora, "F-4G Wild Weasel," p. 117.
41. Price, *War in the Fourth Dimension: US Electronic Warfare from the Vietnam War to the Present*, p. 47.
42. Ibid., p. 5.
43. Ibid.
44. Kendziora, "F-4G Wild Weasel," p. 120.
45. Thornbourough and Mormillo, *Iron Hand: Smashing the Enemy's Air Defenses*, p. 44.
46. USAF, "Tactics and Techniques of Electronic Warfare," p. 40.
47. Kendziora, "F-4G Wild Weasel," p. 6.
48. Price, *War in the Fourth Dimension: US Electronic Warfare from the Vietnam War to the Present*, p. 50.
49. Kendziora, "F-4G Wild Weasel," p. 9.
50. USAF, "Tactics and Techniques of Electronic Warfare," p. 41.
51. Kendziora does not indicate that the F-100F had the capability to carry Shrike; however, another USAF official document indicates that both the F-100F and F-105F fired the missile. Kendziora, "F-4G Wild Weasel"; USAF, "Tactics and Techniques of Electronic Warfare," p. 43.
52. See Thornbourough and Mormillo, *Iron Hand: Smashing the Enemy's Air Defenses*, p. 14; Kendziora, "F-4G Wild Weasel," p. 9.
53. USAF, "Tactics and Techniques of Electronic Warfare," p. 42.
54. Ibid., p. 43.
55. Ibid., p. 44.
56. Ibid., p. 44.
57. Ibid., p. 45.
58. Thornbourough and Mormillo, *Iron Hand: Smashing the Enemy's Air Defenses*, p. 51. Although aircrew were given the option of rotating out of Southeast Asia after 80

missions, most officers took it as a point of pride to attempt the full 100. See Bell, *100 Missions North*.
59. USAF, "Tactics and Techniques of Electronic Warfare," p. 46.
60. Thornbourough and Mormillo, *Iron Hand: Smashing the Enemy's Air Defenses*, p. 69; USAF, "Tactics and Techniques of Electronic Warfare," p. 48.
61. Ibid., pp. 48–49; Thornbourough and Mormillo, *Iron Hand: Smashing the Enemy's Air Defenses*, p. 84.
62. The Air Force Special Communications Center (ASFCC) was the precursor to the Air Force Electronic Warfare Center.
63. Price, *War in the Fourth Dimension: US Electronic Warfare from the Vietnam War to the Present*, pp. 90–91.
64. There were two fighter wings based in Thailand, each at a separate air base, Tikhli and Korat Royal Thai Air Force Bases. Each fighter wing consisted of three squadrons. See Bell, *100 Missions North*.
65. Thornbourough and Mormillo, *Iron Hand: Smashing the Enemy's Air Defenses*, p. 97.
66. USAF, "Tactics and Techniques of Electronic Warfare," p. 20.
67. Ibid., p. 22.
68. Ibid., p. 23.
69. Ibid.
70. Ibid., p. 54.
71. Ibid., p. 27.
72. Thornbourough and Mormillo, *Iron Hand: Smashing the Enemy's Air Defenses*, p. 38.
73. USAF, "Tactics and Techniques of Electronic Warfare," p. 18.
74. Thornbourough and Mormillo, *Iron Hand: Smashing the Enemy's Air Defenses*, p. 40.
75. USAF, "Tactics and Techniques of Electronic Warfare," p. 16.
76. Ibid., p. 55.
77. Ibid.
78. Ibid., p. 56.
79. Ibid., pp. 55–59.
80. Norris, U.S. Air Force Oral History Interview transcript, p. 23.
81. Price, *War in the Fourth Dimension: US Electronic Warfare from the Vietnam War to the Present*, pp. 60–61.
82. Ibid., pp. 59, 74.
83. USAF, "Tactics and Techniques of Electronic Warfare," p. 58.
84. Ibid.
85. Ibid., p. 59.
86. Ibid.
87. Ibid., p. 60.
88. Pape, *Bombing to Win: Air Power and Coercion in War*, p. 197.

89. Ibid., p. 199.
90. Ibid., pp. 201–202.
91. USAF, "Tactics and Techniques of Electronic Warfare," p. 85.
92. See Wayne Thompson, *To Hanoi and Back: The US Air Force and North Vietnam, 1966–1973* (Washington, D.C.: Smithsonian Institution Press, 2000), pp. 219–244.
93. Charles K. Hopkins, Command Historian Eighth Air Force, "Eighth Air Force, 1 July 1972—30 June 1973, Narrative—Part II," Air Force Historical Research Archives, Maxwell Air Force Base, pp. 351–352.
94. Thompson, *To Hanoi and Back: The US Air Force and North Vietnam, 1966–1973*, p. 244.
95. Thornbourough and Mormillo, *Iron Hand: Smashing the Enemy's Air Defenses*, p. 113; USAF, "Tactics and Techniques of Electronic Warfare," p. 6.
96. Thornbourough and Mormillo, *Iron Hand: Smashing the Enemy's Air Defenses*, p. 103.
97. Ibid., p. 131.
98. Hopkins, "Eighth Air Force, 1 July 1972—30 June 1973, Narrative—Part II," p. 409.
99. USAF, "Tactics and Techniques of Electronic Warfare," pp. 85–86.
100. Ibid., pp. 89–90.
101. Ibid., p. 90.
102. Thornbourough and Mormillo, *Iron Hand: Smashing the Enemy's Air Defenses*, p. 123.
103. Ibid., p. 119.
104. Hopkins, "Eighth Air Force, 1 July 1972—30 June 1973, Narrative—Part II," p. 409.
105. Ibid.
106. Ibid., p. 410.
107. Major Richard J. Smith, interview by Richard Emmons and Hugh Ahman, June 20–21, 1974, CD recording, Air Force Historical Research Archives, Maxwell Air Force Base.
108. Kendziora, "F-4G Wild Weasel," p. 2.
109. USAF, "Tactics and Techniques of Electronic Warfare," p. 57.
110. Ibid., p. 88.
111. Derived from table in Thompson, *To Hanoi and Back: The US Air Force and North Vietnam, 1966–1973*, p. 311. Thompson does not distinguish between losses incurred over South Vietnam and those over North Vietnam in this table.
112. See Hopkins, "Eighth Air Force, 1 July 1972—30 June 1973, Narrative—Part II."
113. Thornbourough and Mormillo, *Iron Hand: Smashing the Enemy's Air Defenses*, p. 140.
114. While laser- and television-guided bombs made their appearance in the Vietnam War, they came too late to have an effect during Operation Rolling Thunder. For a discussion of laser-guided bomb use in Vietnam, see Drew Middleton, *Air War—Vietnam* (New York: Arno Press, 1978), pp. 79–92.

115. Thomas C. Hone, "Southeast Asia," in Benjamin Franklin Cooling, ed., *Case Studies in Air Superiority. Special Studies* (Washington, D.C.: Center for Air Force History, 1994), pp. 506–507.
116. While the USN lacked a dedicated Wild Weasel aircraft, it used similar tactics as the USAF for SEAD. Lon O. Nordeen, *Air Warfare in the Missile Age*, 2nd ed. (Washington, DC: Smithsonian Institution Press, 2002), p. 38.

Chapter 5

1. A "task force" is a group of naval vessels organized with a particular mission in mind. In this case, the task was to enable an amphibious landing on the Falkland Islands in order to regain possession on behalf of the United Kingdom. This task also implied that it was necessary to hold at bay Argentinian air and naval forces long enough to accomplish retaking the islands.
2. CIWS is a generic term encompassing systems which incorporate a high rate of fire gun, self-contained radar and fire control, and automated operation. David Miller and Chris Miller, *Modern Naval Combat* (New York: Crescent Books, 1986), p. 197. Later examples also include short-range missile systems, such as the RIM-116 Rolling Airframe Missile (RAM) in U.S. and Allied service. "RIM-116 Rolling Airframe Missile (RAM)" www.navy.mil/Resources/Fact-Files/Display-FactFiles/Article/2168961/rim-116-rolling-airframe-missile-ram/, accessed May 20, 2020.
3. Alastair Finlan, *The Royal Navy in the Falklands Conflict and the Gulf War: Culture and Strategy* (London, Frank Cass, 2004), p. 57.
4. Martin Middlebrook, *Operation Corporate: The Falklands War, 1982* (London: Viking, 1985), p. 133.
5. David Hart Dyke, *Four Weeks in May: The Loss of the HMS* Coventry, *A Captain's Story* (London: Atlantic Books, 2007), p. 137; D. George Boyce, *The Falklands War* (New York: Palgrave Macmillan, 2005), p. 74.
6. David Brown, *The Royal Navy and the Falklands War* (London: Leo Cooper Ltd., 1987), p. 114.
7. Phillip D. Grove, "Falklands Conflict 1982—The Air War: A New Appraisal," in Stephen Badsey, Rob Havers, and Eric Grove, eds., *The Falklands Islands Conflict Twenty Years On: Lessons for the Future*, (London: Frank Cass, 2005), p. 269; Boyce, *The Falklands War*, p. 74.
8. Ruben O. Moro, *The History of the South Atlantic Conflict: The War for the Malvinas* (New York: Praeger, 1989), pp. 78–80; C.W. Koburger, "Glory Manque: Argentina in the Falklands," *Navy International*, Vol. 88, No. 5 (May 1983), p. 265; Robert L. Scheina, "Argentina's Navy in the Falklands War," in *Jane's Naval Review, 1983–84* (London: Jane's, 1983–1987), p. 22.
9. Anonymous, "Falklands War: 21st May," in *Warplane*, No. 44 (1986), p. 864.

10. Jeff Ethell and Michael O'Leary, "Argentine air war: the other side," in *Air Progress*, Vol. 45, No. 5, p. 28.
11. Derived from Brown, *The Royal Navy and the Falklands War*, pp. 371–372. It is unknown how practicable coordinating a near-simultaneous launch from every Exocet-capable ship in the Argentinian Navy would have been. However, if the Navy had managed a salvo of just fifteen missiles—less than half of its launch capacity on paper—it would have tripled its total air-launched arsenal.
12. Ethell and O'Leary, "Argentine air war: the other side," p. 33.
13. Salvador Mafe Huertas, "The other side of the coin," in *Air International*, Vol. 24, No. 5 (May 1983), pp. 216–217.
14. Scheina, "Argentina's Navy in the Falklands War," p. 27; Brown, *The Royal Navy and the Falklands War*, pp. 113–117; Moro, *The History of the South Atlantic Conflict: The War for the Malvinas*, p. 79; Sandy Woodward, *One Hundred Days: the Memoirs of the Falklands Battle Group Commander* (London: Harper Collins, 1992), p. 175; Brigadier Horacio Mir Gonzalez, "An Argentine Airman in the South Atlantic," in Badsey, Havers, and Grove, *The Falklands Islands Conflict Twenty Years On: Lessons for the Future*, p. 76.
15. Ethell and O'Leary, "Argentine air war: the other side," p. 37.
16. Eric J. Grove, *Vanguard to Trident: British Naval Policy Since World War II* (Annapolis: Naval Institute Press, 1987), pp. 245–250.
17. See Emily O. Goldman, "Receptivity to Revolution" in Goldman and Eliason, pp. 277–280.
18. Grove, pp. 255–256, 275, 310–311. The story of post-war British defense policy is of course a complex one, and I have painted the picture in the broadest of brushstrokes. This is justifiable, however, because the primary reason for this discussion is not the how and the why of defense policy, and the BRN in particular, but the state of the Navy in 1982. To go into more detail would simply duplicate Grove's work here.
19. Finlan, *The Royal Navy in the Falklands Conflict and the Gulf War: Culture and Strategy*, p. 38; United Kingdom, Secretary of State for Defense, *The Statement on Defense Estimates* (London: Her Majesty's Stationary Office, 1981), p. 25.
20. Boyce, *The Falklands War*, p. 71.
21. Ibid., p. 72.
22. Anthony Preston, *Sea Combat Off the Falklands: The Lessons That Must Be Learned* (London: Willow Books, 1982), p. 11.
23. Ibid.
24. With one squadron of five aircraft on each carrier, the two carriers would normally deploy ten SHAR. Reinforced with additional aircraft and absorbing the training squadron brought the total number of aircraft in the task force to twenty. Brown, *The Royal Navy and the Falklands War*, p. 69.
25. Koburger, "Glory Manque: Argentina in the Falklands," p. 262.
26. Max Hastings and Simon Jenkins, *The Battle for the Falklands* (New York: W.W. Norton and Company, 1983), p. 48.

27. See Middlebrook, *Operation Corporate: The Falklands War, 1982*, pp. 40–52.
28. Koburger, "Glory Manque: Argentina in the Falklands," p. 265.
29. Salvador Mafe Huertas, "South Atlantic Air War: The other side of the coin," in *Air International*, Vol. 24, No. 5 (May 1983), pp. 216–217.
30. Boyce, *The Falklands War*, p. 75.
31. Ibid., p. 70.
32. The BRN leadership was in fact one of the prime advocates of sending the task force in response to the invasion. See Middlebrook, *Operation Corporate: The Falklands War, 1982*, pp. 67–69; Hastings and Jenkins, *The Battle for the Falklands*, pp. 61–70.
33. "Area Defense" weapons refer to medium- to long-range missiles that enable a ship to engage threats to other ships, in addition to protecting itself. In the early 1980s, the typical range for these types of missiles was 30–40 miles and they could reach altitudes of about 80,000 feet. Examples besides Sea Dart include the U.S. Standard Missile (SM-1, SM-2), and the Soviet S-300F and M-11 (which NATO code-named SA-N-6 and SA-N-4, respectively).

"Point Defense" weapons are designed as the last line of defense against air threats (missiles or aircraft). Their range is typically only a few miles. Given the speed of modern threats (in the early 1980s as well as today), effective point defense often requires automation. Examples include the U.S. "Phalanx" CIWS gun system and British Sea Wolf missile. The Soviets developed a system similar to Phalanx.
34. Grove, p. 317.
35. Brown, *The Royal Navy and the Falklands War*, p. 234.
36. Preston, *Sea Combat Off the Falklands: The Lessons That Must Be Learned*, p. 22.
37. Ibid., p. 25.
38. Finlan, *The Royal Navy in the Falklands Conflict and the Gulf War: Culture and Strategy*, p. 56.
39. Admiral Sir Jeremy Black, *There and Back* (London: Elliot and Thompson, 2005), p. 165.
40. Brown, *The Royal Navy and the Falklands War*, p. 160.
41. Ibid., p. 81.
42. Roy Braybrook, "Sea Harriers in the South Atlantic," p. 42.
43. Joseph Udemi, "Modified to fit the need: British aircraft in the Falklands," in *Airpower Journal*, Spring 1989, pp. 56–57.
44. Ibid., p. 58.
45. Nigel D. Ward, "Sea Harrier Over the Falklands: A Maverick at War," unpublished manuscript, November 1991, manuscript collection, Imperial War Museum, London; Udemi, "Modified to fit the need: British aircraft in the Falklands," p. 53.
46. Ibid., p. 61.
47. Ibid., p. 52.
48. Henry Leach, *Endure No Makeshifts: Some Naval Recollections* (London: Leo Cooper, 1993), p. 201.
49. Brown, *The Royal Navy and the Falklands War*, p. 65.

50. Ibid., pp. 65, 96.
51. Ibid., p. 75.
52. Ibid., pp. 365–370.
53. Black, *There and Back*, p. 153.
54. Canopis (pseudonym), "A personal view of the Falklands campaign," in *Naval Review*, Vol. 71, No. 1 (January 1983), p. 19.
55. Boyce, *The Falklands War*, p. 74.
56. Dyke, *Four Weeks in May: The Loss of the HMS* Coventry, *A Captain's Story*, p. 88.
57. T.D. Cairns, ed., "HMS *Invincible*: some individual recollections," in *Naval Review*, Vol. 71 No. 2 (April 1983), p. 113.
58. Black, *There and Back*, p. 165.
59. The Neptune is a twin-engine propeller driven aircraft equipped with surveillance radar.
60. Dyke, *Four Weeks in May: The Loss of the HMS* Coventry, *A Captain's Story*, p. 192; Brown, *The Royal Navy and the Falklands War*, p. 141; Miller and Miller, *Modern Naval Combat*, p. 197.
61. Michael Thomas Nicholson, ITN correspondent, in Hugh McManners, *Forgotten Voices of the Falklands* (London: Ebury Press in association with the Imperial War Museum, 2008), with introduction by Sir Max Hastings, p. 166. This work is a collection of direct quotations from interviews of participants in the war, and as such is very close to a primary source document, even though it is a published work.
62. Brown, *The Royal Navy and the Falklands War*, p. 144; Middlebrook, *Operation Corporate: The Falklands War, 1982*, pp. 155–157.
63. Brown, *The Royal Navy and the Falklands War*, p. 141.
64. Ethell and O'Leary, "Argentine air war: the other side," p. 31.
65. Brown, *The Royal Navy and the Falklands War*, p. 145.
66. Ethell and O'Leary, "Argentine air war: the other side," p. 228.
67. Boyce, *The Falklands War*, p. 123.
68. Black, *There and Back*, p. 166.
69. Boyce, *The Falklands War*, p. 63.
70. Black, *There and Back*, p. 156.
71. Commander Nigel David 'Sharkey' Ward, Commander 801 Squadron, in McManners, *Forgotten Voices of the Falklands*, p. 212.
72. Ward, 149.
73. Boyce, *The Falklands War*, p. 64.
74. Miller and Miller, *Modern Naval Combat*, pp. 194–195.
75. Boyce, *The Falklands War*, p. 71; 98; 110.
76. Quoted in Woodward, *One Hundred Days: the Memoirs of the Falklands Battle Group Commander*, p. 80.
77. Ibid., p. 71.
78. Dyke, *Four Weeks in May: The Loss of the HMS* Coventry, *A Captain's Story*, p. 92.

79. Captain Jeremy Larken, HMS *Fearless*, in McManners, *Forgotten Voices of the Falklands*, p. 216.
80. Ibid., p. 119.
81. For a full account see Brown, *The Royal Navy and the Falklands War*, pp. 186–227; Middlebrook, *Operation Corporate: The Falklands War, 1982*, pp. 206–237; Hastings and Jenkins, *The Battle for the Falklands*, pp. 200–232.
82. Koburger, "Glory Manque: Argentina in the Falklands," p. 266.
83. Canopis (psuedonym), p. 19.
84. Boyce, *The Falklands War*, p. 122.
85. Lieutenant Commander Graham John Edmonds, Operations Officer HMS *Broadsword*, in McManners, *Forgotten Voices of the Falklands*, p. 325.
86. Robert S. Bolia, "The Falklands War: The Bluff Cove Disaster," *Military Review* (November—December 2004), p. 71.
87. Boyce, *The Falklands War*, p. 69; Major Simon Ewen Southby-Tailyour, Royal Marines, in McManners, *Forgotten Voices of the Falklands*, p. 332.
88. Dyke, *Four Weeks in May: The Loss of the HMS* Coventry, *A Captain's Story*, p. 148.
89. Ibid.
90. Radio Supervisor Stewart Anthony MacFarlane, HMS *Coventry*, in McManners, *Forgotten Voices of the Falklands*, pp. 220–221.
91. Mason, p. 39.
92. Mason, p. 37.
93. Scheina, "Argentina's Navy in the Falklands War," p. 27.
94. Koburger, "Glory Manque: Argentina in the Falklands," p. 265.
95. Ibid., p. 267.
96. Boyce, *The Falklands War*, p. 73.
97. The BRN used three torpedoes of a design older than its 1938-vintage target for this result. William Ruhe, "Submarine Lessons," in Bruce Watson and Peter Dunn, eds., *Military Lessons of the Falkland Islands War: Views From the United States* (London: Arms and Armor Press, 1984), p. 9.
98. Boyce, *The Falklands War*, p. 123.
99. Dyke, *Four Weeks in May: The Loss of the HMS* Coventry, *A Captain's Story*, p. 110.

Chapter 6

1. Max Hastings, *Retribution: The Battle for Japan, 1944–45* (New York: Vintage, 2007), p. 266.
2. Ibid., p. 279.
3. Bernard Ireland and Eric Grove, *Jane's War at Sea, 1897–1997: 100 Years of Jane's Fighting Ships* (New York: Harper Collins, 1997), p. 158; and Hastings, *Retribution: The Battle for Japan, 1944–45*, p. 268.

4. For a discussion of Japan's imperial ambitions, see Jack Snyder, *Myths of Empire: Domestic Politics and International Ambition* (Ithaca: Cornell University Press, 1991), p. 112–152.
5. The Washington Naval Treaty regime set limits on the size of the navies of the United States, Great Britain, Japan, France, and Italy. It had been proposed by the Warren G. Harding administration in order to avoid a naval arms competition between the Great Powers that the U.S. Congress would not fund. By establishing ratios of capital ships between these powers, the treaty forestalled a naval arms race until the regime fell apart in the mid-1930s. The ratio "5:5:3:1:1" meant that the U.S. and Great Britain could have five battleships to Japan's three, and France and Italy's one. The treaty had the unintended consequence of increased aircraft carrier production, as some capital ship hulls already in production were re-purposed. Japan built up its fleet to the limits of treaty restrictions, while economic hardship from the Great Depression hindered the United States from doing the same. The result was that when the treaty lapsed, Japan had a more favorable ratio of capital ships to that of the United States—approximately ten U.S. ships to seven Japanese versus the treaty ratio of 5:3. See George W. Baer, *One Hundred Years of Sea Power: The U.S. Navy, 1890–1990* (Stanford: Stanford University Press, 1994), pp. 94–101, 117–118. For a discussion of the follow-on London Naval Conference, see John H. Maurer and Christopher M. Bell, eds., *At the Crossroads Between Peace and War: The London Naval Conference of 1930* (Annapolis: Naval Institute Press, 2014).
6. David C. Evans and Mark R. Peattie, *Kaigun: Strategy, Tactics, and Technology in the Imperial Japanese Navy, 1887–1941* (Annapolis: Naval Institute Press, 1997), pp. 353–354.
7. Ibid., pp. 357–359.
8. Ibid., p. 360.
9. Ratios calculated from Table 10-3, Evans and Peattie, *Kaigun: Strategy, Tactics, and Technology in the Imperial Japanese Navy, 1887–1941*, p. 365.
10. Evans and Peattie, *Kaigun: Strategy, Tactics, and Technology in the Imperial Japanese Navy, 1887–1941*, p. 439.
11. Ibid., p. 383.
12. Ibid., p. 425.
13. Ibid., p. 430.
14. Ibid., p. 435.
15. Ibid., pp. 434–435.
16. Ibid., pp. 430–431.
17. Ibid., p. 437.
18. Ibid., p. 441.
19. Ibid., p. 436.
20. Ibid., p. 438.
21. Ibid., p. 437.

22. Clay Blair Jr., *Silent Victory: The US Submarine War Against Japan* (Annapolis: Naval Institute Press, 2001), p. 106.
23. Hastings, *Retribution: The Battle for Japan, 1944–45*, p. 269.
24. Blair, *Silent Victory: The US Submarine War Against Japan*, p. 361.
25. Ibid., p. 360
26. Ibid., p. 360.
27. Ibid., p. 361.
28. Ibid., p. 359.
29. Ibid., p. 362.
30. Ibid., p. 360.
31. Ibid., p. 551.
32. Ibid., p. 552.
33. Ibid., p. 554.
34. Ibid., p. 553.
35. Ibid., p. 554.
36. Hastings, *Retribution: The Battle for Japan, 1944–45*, p. 273.
37. Blair, *Silent Victory: The US Submarine War Against Japan*, pp. 616–617.
38. Ibid., pp. 616–618.
39. Ibid., pp. 818–819.
40. Submarine Force, Pacific Fleet, U.S. Navy, "Submarine Operational History of World War II," May 14, 1947, Vol. 1, U.S. Submarine War Patrol Reports and Related Documents, 1941–1945, Brandeis University Microfilm Archives, Waltham, Mass., p. 280.
41. Blair, *Silent Victory: The US Submarine War Against Japan*, p. 855.
42. Ibid., pp. 859–864.
43. Ibid., p. 878.
44. Evans and Peattie, *Kaigun: Strategy, Tactics, and Technology in the Imperial Japanese Navy, 1887–1941*, p. 496.
45. Blair, *Silent Victory: The US Submarine War Against Japan*, p. 879.
46. Ibid., p. 881.
47. Bureau of Ships Preliminary Design Branch, "Submarine Report: Depth charge, bomb, mine, torpedo and gunfire damage including losses in action, 7 December 1941 to 15 August 1945," January 1, 1949, U.S. Submarine War Patrol Reports and Related Documents, 1941–1945, Brandeis University Microfilm Archives, Waltham, Mass., p. 6.
48. Hastings, *Retribution: The Battle for Japan, 1944–45*, p. 269.
49. Bureau of Ships, January 1, 1949, p. 8.
50. Submarine Force, Pacific Fleet, U.S. Navy, "Submarine Operational History of World War II," May 14, 1947, Vol. 1, U.S. Submarine War Patrol Reports and Related Documents, 1941–1945, Brandeis University Microfilm Archives, Waltham, Mass., p. 242.
51. Submarine Force, Pacific Fleet, May 14, 1947, pp. 271–272.

52. Ireland and Grove, *Jane's War at Sea, 1897–1997: 100 Years of Jane's Fighting Ships*, p. 161.

53. Evans and Peattie, *Kaigun: Strategy, Tactics, and Technology in the Imperial Japanese Navy, 1887–1941*, pp. 364–366.

54. Submarine Force Pacific Fleet, May 14, 1947, p. 257.

55. Ibid., pp. 268–269.

56. Ibid., p. 263.

57. Bureau of Ships, January 1, 1949, p. 10; Submarine Force Pacific Fleet, May 14, 1947, p. 245.

58. Bureau of Ships, January 1, 1949, p. 10.

59. Ibid.

60. Ibid., p. 8; Submarine Force, Pacific Fleet, May 14, 1947, p. 243.

61. Hastings, *Retribution: The Battle for Japan, 1944–45*, p. 270.

62. Naval Analysis Division, United States Strategic Bombing Survey (Pacific), NAV No. 11, p. 59; Submarine Force, Pacific Fleet, May 14, 1947, p. 242.

63. Submarine Force Pacific Fleet, May 14, 1947, p. 245

64. Naval Analysis Division, USSBS (Pacific), NAV No. 11, p. 59.

65. Bureau of Ships, January 1, 1949, p. 10.

66. Ibid., p. 11.

67. Ibid., p. 12.

68. Ibid.

69. Bureau of Ships, January 1, 1949, pp. 15–16.

70. Ibid., p. 16.

71. Submarine Force, Pacific Fleet, May 14, 1947, pp. 257–259.

72. Ibid., p. 260.

73. Bureau of Ships, January 1, 1949, p. 9.

74. Naval Analysis Division, USSBS (Pacific), Interrogation of LCDR Noriteru Yatsui IJN, "Attacks on Japanese Shipping," NAV No. 37, p. 161.

75. Submarine Force, Pacific Fleet, May 14, 1947, p. 274.

76. Bureau of Ships, January 1, 1949, p. 9; Submarine Force Pacific Fleet, May 14, 1947, p. 247; 274; Naval Analysis Division, USSBS (Pacific), Interrogation of LCDR T. Okamoto IJN "Japanese Airborne Magnetic Detector," NAV No. 48, p. 197.

77. Naval Analysis Division, USSBS (Pacific), NAV No. 11, p. 59.

78. Submarine Force, Pacific Fleet, May 14, 1947, pp. 274–275; Naval Analysis Division, USSBS (Pacific), NAV No. 48, p. 197.

79. Submarine Force, Pacific Fleet, May 14, 1947, p. 273.

80. Naval Analysis Division, USSBS (Pacific), NAV No. 47, pp. 195–196.

81. Bureau of Ships, January 1, 1949, p. 9.

82. Ibid.

83. Naval Analysis Division, USSBS (Pacific), Interrogation of LCDR T. Okamoto IJN, "Antisubmarine Training and Equipment," NAV No. 63, p. 259.

84. Submarine Force, Pacific Fleet, May 14, 1947, p. 275; Naval Analysis Division, USSBS (Pacific), Interrogation of LT (jg) Y Okuno IJN, "Aircraft Escort of Convoys and Anti-Submarine Operations," NAV No. 56, p. 228; Naval Analysis Division, USSBS (Pacific), NAV No. 11, p. 58.
85. See Ireland and Grove, *Jane's War at Sea, 1897–1997: 100 Years of Jane's Fighting Ships*, p. 156.
86. Submarine Force, Pacific Fleet, May 14, 1947, p. 275.
87. Bureau of Ships, January 1, 1949, pp. 13–14; Naval Analysis Division, USSBS (Pacific), NAV No. 63, p. 260.
88. Submarine Force, Pacific Fleet, May 14, 1947, p. 273.
89. Ibid.; Naval Analysis Division, USSBS (Pacific), NAV No. 11, p. 59.
90. Naval Analysis Division, USSBS (Pacific), Interrogation of LCDR Shisei Yagumoto, IJN, "Escort and Protection of Shipping," NAV No. 45, pp. 190, 195.
91. Naval Analysis Division, USSBS (Pacific), NAV No. 11, p. 59.
92. Naval Analysis Division, USSBS (Pacific), NAV No. 45, p. 195.
93. Naval Analysis Division, United States Strategic Bombing Survey (Pacific), Interrogation of CAPT Atsushi Oi, IJN, "Japanese Naval Escort of Shipping and Shipping Losses," NAV No. 11, in *Interrogations of Japanese Officials*, OPNAV-P-03–100, Vol. I, p. 56.
94. Evans and Peattie, *Kaigun: Strategy, Tactics, and Technology in the Imperial Japanese Navy, 1887–1941*, p. 438.
95. Naval Analysis Division, USSBS (Pacific), NAV No. 11, p. 56.
96. Submarine Force, Pacific Fleet, May 14, 1947, p. 262.
97. Ibid.
98. Ireland and Grove, *Jane's War at Sea, 1897–1997: 100 Years of Jane's Fighting Ships*, p. 161.
99. Submarine Force, Pacific Fleet, May 14, 1947, p. 270. The 901st Air Squadron is also sometimes referred to as 901 Air Fleet. Its headquarters was established at Tateyama, Honshu, Japan.
100. Naval Analysis Division, USSBS (Pacific), NAV No. 37, p. 161.
101. Naval Analysis Division, USSBS (Pacific), NAV No. 47, p. 194; NAV No. 45, p. 186;
102. Submarine Force, Pacific Fleet, May 14, 1947, pp. 272–273.
103. Naval Analysis Division, USSBS (Pacific), Interrogation of CDR Masataka Chihaya IJN, "Convoy Protection of Shipping: Netherlands East Indies—New Guinea," NAV No. 49, p. 201.
104. This was the 4th Advanced Southern Fleet, headquartered at Ambon. Interrogation of CDR Masataka Chihaya IJN, "Convoy Protection of Shipping: Netherlands East Indies—New Guinea," NAV No. 49, p. 201.
105. Naval Analysis Division, USSBS (Pacific), NAV No. 11, pp. 57–58.
106. Bureau of Ships, January 1, 1949, pp. 9–10.
107. Submarine Force Pacific Fleet, May 14, 1947, p. 249.

108. Ibid.
109. Ibid., pp. 249–251.
110. Ireland and Grove, *Jane's War at Sea, 1897–1997: 100 Years of Jane's Fighting Ships*, p. 156.
111. Ibid., p. 157.
112. Submarine Force Pacific Fleet, May 14, 1947, p. 277.
113. Ibid., p. 280.
114. The reader will recall the BRN's *de facto* organization as consisting of the Grand Fleet and "everything else."
115. The reader will note the aforementioned U-boat offensive of World War I. Also, the U-boat offensive of World War II predated the U.S. offensive.
116. For example, U.S. submarines increased their vulnerability to detection when having to fire multiple salvoes of torpedoes at the same target because of the high incidence of duds.
117. Harvey M. Sapolsky, *The Polaris System Development: Bureaucratic and Programmatic Success in Government* (Cambridge, Mass.: Harvard University Press, 1972), pp. 44, 59.

Chapter 7

1. See Meir Finkel, *On Flexibility: Recovery from Technological and Doctrinal Surprise on the Battlefield* (Stanford: Stanford University Press, 2001), p. 2.

About the Author

Matthew Tattar is an Associate Professor in the War Gaming Department, Center for Naval Warfare Studies at the U.S. Naval War College. His research focuses on military innovation and strategic deterrence. Previously, he was an International Security Fellow at the Belfer Center for Science and International Affairs at Harvard Kennedy School. He holds a PhD from Brandeis University.

Index

Belfer Center Studies in International Security

Published by The MIT Press
Steven E. Miller and Jacqueline L. Hazelton, series editors
Karen Motley, executive editor
Belfer Center for Science and International Affairs
Harvard Kennedy School, Harvard University

Acharya, Amitav, and Evelyn Goh, eds., *Reassessing Security Cooperation in the Asia-Pacific* (2007)

Agha, Hussein, Shai Feldman, Ahmad Khalidi, and Zeev Schiff, *Track-II Diplomacy: Lessons from the Middle East* (2003)

Allison, Graham, and Robert D. Blackwill, with Ali Wyne, *Lee Kuan Yew: The Grand Master's Insights on China, the United States, and the World* (2012)

Allison, Graham T., Owen R. Coté, Jr., Richard A. Falkenrath, and Steven E. Miller, *Avoiding Nuclear Anarchy: Containing the Threat of Loose Russian Nuclear Weapons and Fissile Material* (1996)

Allison, Graham T., and Kalypso Nicolaïdis, eds., *The Greek Paradox: Promise vs. Performance* (1996)

Arbatov, Alexei, Abram Chayes, Antonia Handler Chayes, and Lara Olson, eds., *Managing Conflict in the Former Soviet Union: Russian and American Perspectives* (1997)

Bennett, Andrew, *Condemned to Repetition? The Rise, Fall, and Reprise of Soviet-Russian Military Interventionism, 1973–1996* (1999)

Blackwill, Robert D., and Michael Stürmer, eds., *Allies Divided: Transatlantic Policies for the Greater Middle East* (1997)

Blackwill, Robert D., and Paul Dibb, eds., *America's Asian Alliances* (2000)

Blum, Gabriella, and Philip B. Heymann, *Laws, Outlaws, and Terrorists: Lessons from the War on Terrorism* (2010)

Brom, Shlomo, and Yiftah Shapir, eds., *The Middle East Military Balance 1999–2000* (1999)

Brom, Shlomo, and Yiftah Shapir, eds., *The Middle East Military Balance 2001–2002* (2002)

Brown, Michael E., ed., *The International Dimensions of Internal Conflict* (1996)

Brown, Michael E., and Šumit Ganguly, eds., *Fighting Words: Language Policy and Ethnic Relations in Asia* (2003)

Brown, Michael E., and Šumit Ganguly, eds., *Government Policies and Ethnic Relations in Asia and the Pacific* (1997)

Carter, Ashton B., and John P. White, eds., *Keeping the Edge: Managing Defense for the Future* (2001)

Chenoweth, Erica, and Adria Lawrence, eds., *Rethinking Violence: State and Non-State Actors in Conflict* (2010)

de Nevers, Renée, *Comrades No More: The Seeds of Political Change in Eastern Europe* (2003)

Elman, Colin, and Miriam Fendius Elman, eds., *Bridges and Boundaries: Historians, Political Scientists, and the Study of International Relations* (2001)

Elman, Colin, and Miriam Fendius Elman, eds., *Progress in International Relations Theory: Appraising the Field* (2003)

Elman, Miriam Fendius, ed., *Paths to Peace: Is Democracy the Answer?* (1997)

Falkenrath, Richard A., *Shaping Europe's Military Order: The Origins and Consequences of the CFE Treaty* (1994)

Falkenrath, Richard A., Robert D. Newman, and Bradley A. Thayer, *America's Achilles' Heel: Nuclear, Biological, and Chemical Terrorism and Covert Attack* (1998)

Feaver, Peter D., and Richard H. Kohn, eds., *Soldiers and Civilians: The Civil-Military Gap and American National Security* (2001)

Feldman, Shai, *Nuclear Weapons and Arms Control in the Middle East* (1996)

Feldman, Shai, and Yiftah Shapir, eds., *The Middle East Military Balance 2000–2001* (2001)

Findlay, Trevor, *Transforming Nuclear Safeguards Culture: The IAEA, Iraq, and the Future of Non-Proliferation* (2022)

Forsberg, Randall, ed., *The Arms Production Dilemma: Contraction and Restraint in the World Combat Aircraft Industry* (1994)

George, Alexander L., and Andrew Bennett, *Case Studies and Theory Development in the Social Sciences* (2005)

Gilroy, Curtis, and Cindy Williams, eds., *Service to Country: Personnel Policy and the Transformation of Western Militaries* (2007)

Hagerty, Devin T., *The Consequences of Nuclear Proliferation: Lessons from South Asia* (1998)

Heymann, Philip B., *Terrorism and America: A Commonsense Strategy for a Democratic Society* (1998)

Heymann, Philip B., *Terrorism, Freedom, and Security: Winning without War* (2003)

Heymann, Philip B., and Juliette N. Kayyem, *Protecting Liberty in an Age of Terror* (2005)

Howitt, Arnold M., and Robyn L. Pangi, eds., *Countering Terrorism: Dimensions of Preparedness* (2003)

Hudson, Valerie M., and Andrea M. den Boer, *Bare Branches: The Security Implications of Asia's Surplus Male Population* (2004)

Kayyem, Juliette N., and Robyn L. Pangi, eds., *First to Arrive: State and Local Responses to Terrorism* (2003)

Kokoshin, Andrei A., *Soviet Strategic Thought, 1917–91* (1998)

Lawson, Chappell, Alan Bersin, and Juliette Kayyem, eds., *Beyond 9/11: Homeland Security for the Twenty-First Century* (2020)

Lederberg, Joshua, ed., *Biological Weapons: Limiting the Threat* (1999)

Mansfield, Edward D., and Jack Snyder, *Electing to Fight: Why Emerging Democracies Go to War* (2005)

Martin, Lenore G., and Dimitris Keridis, eds., *The Future of Turkish Foreign Policy* (2004)

May, Ernest R., and Philip D. Zelikow, eds., *Dealing with Dictators: Dilemmas of U.S. Diplomacy and Intelligence Analysis, 1945–1990* (2007)

Phillips, David L., *Liberating Kosovo: Coercive Diplomacy and U.S. Intervention* (2012)

Poneman, Daniel B., *Double Jeopardy: Combating Nuclear Terror and Climate Change* (2019)

Rosecrance, Richard N., and Steven E. Miller, eds., *The Next Great War? The Roots of World War I and the Risk of U.S.-China Conflict* (2015)

Shaffer, Brenda, *Borders and Brethren: Iran and the Challenge of Azerbaijani Identity* (2002)

Shaffer, Brenda, ed., *The Limits of Culture: Islam and Foreign Policy* (2006)

Shields, John M., and William C. Potter, eds., *Dismantling the Cold War: U.S. and NIS Perspectives on the Nunn-Lugar Cooperative Threat Reduction Program* (1997)

Tattar, Matthew A., *Innovation and Adaptation in War* (2024)

Tucker, Jonathan B., ed., *Toxic Terror: Assessing Terrorist Use of Chemical and Biological Weapons* (2000)

Utgoff, Victor A., ed., *The Coming Crisis: Nuclear Proliferation, U.S. Interests, and World Order* (2000)

Weiner, Sharon K., *Our Own Worst Enemy? Institutional Interests and the Proliferation of Nuclear Weapons Expertise* (2011)

Williams, Cindy, ed., *Filling the Ranks: Transforming the U.S. Military Personnel System* (2004)

Williams, Cindy, ed., *Holding the Line: U.S. Defense Alternatives for the Early 21st Century* (2001)

Xu Qiyu, *Fragile Rise: Grand Strategy and the Fate of Imperial Germany, 1871–1914*, trans. Joshua Hill (2017)

Zoughbie, Daniel E., *Indecision Points: George W. Bush and the Israeli-Palestinian Conflict* (2014)

The Belfer Center for Science and International Affairs

Meghan L. O'Sullivan, Director
Harvard Kennedy School
79 JFK Street, Cambridge, MA 02138
Tel: (617) 495-1400 | Fax: (617) 495-8963
http://www.belfercenter.org | belfer_center@hks.harvard.edu

The Belfer Center is the hub of Harvard Kennedy School's research, teaching, and training in international security affairs, environmental and resource issues, and science and technology policy.

The Center has a dual mission: (1) to provide leadership in advancing policy-relevant knowledge about the most important challenges of international security and other critical issues where science, technology, environmental policy, and international affairs intersect; and (2) to prepare future generations of leaders for these arenas. Center researchers not only conduct scholarly research, but also develop prescriptions for policy reform. Faculty and fellows analyze global challenges from nuclear proliferation and terrorism to climate change and energy policy.

The Belfer Center's leadership begins with the recognition of science and technology as driving forces constantly transforming both the challenges we face and the opportunities for problem-solving. Building on the vision of founder Paul Doty, the Center addresses serious global concerns by integrating insights and research of social scientists, natural scientists, technologists, and practitioners in government, diplomacy, the military, and business.

The heart of the Belfer Center is its resident research community of more than 150 scholars, including Harvard faculty, researchers, practitioners, and each year a new, international, interdisciplinary group of research fellows. Through publications and policy discussions, workshops, seminars, and conferences, the Center promotes innovative solutions to significant national and international challenges.

The Center's International Security Program, directed by Steven E. Miller, sponsors and edits the Belfer Center Studies in International Security and the quarterly journal *International Security*.

The Center is supported by an endowment established with funds from Robert and Reneé Belfer, the Ford Foundation, and Harvard University, by foundation grants, by individual gifts, and by occasional government contracts.

Publisher contact:
The MIT Press
Massachusetts Institute of Technology
77 Massachusetts Avenue, Cambridge, MA 02139
mitpress.mit.edu

EU Authorised Representative:
Easy Access System Europe, Mustamäe tee 50,
10621 Tallinn, Estonia
gpsr.requests@easproject.com

Printed by Integrated Books International,
United States of America